The Vicarious Brain, Creator of Worlds

THE VICARIOUS BRAIN,

Creator of Worlds

ALAIN BERTHOZ

Translated by Giselle Weiss

HARVARD UNIVERSITY PRESS

Cambridge, Massachusetts, and London, England   2017

First printing

First published as *La Vicariance: Le cerveau créateur de mondes*
© Odile Jacob, 2013

Library of Congress Cataloging-in-Publication Data

Names: Berthoz, A., author.
Title: The vicarious brain, creator of worlds/Alain Berthoz ; translated by
    Giselle Weiss.
Other titles: Vicariance English
Description: Cambridge, Massachusetts : Harvard University Press, 2017. |
    "First published as La Vicariance: Le cerveau créateur de mondes (c) Odile
    Jacob, 2013."—Title page verso | Includes bibliographical references and index.
Identifiers: LCCN 2016022571 | ISBN 9780674088955 (alk. paper)
Subjects: LCSH: Adaptability (Psychology) | Adjustment (Psychology) | Mind
    and reality. | Brain. | Identity (Psychology) | Avatars (Virtual reality)
Classification: LCC BF335 .B4713 2017 | DDC 153—dc23 LC record available
    at https://lccn.loc.gov/2016022571

For Catherine

Contents

Part IV: Education

Prologue

"Only the hand that erases." This very strong, very Latin verb conjures up the *rasura,* the gesture that erases and liberates, so that a new text, a new story may start, so that the hand should be free to create, without having to wrestle with the scribbles in the margins, so that this whiteness should be a new beginning. This demanding radicality in search of an absolute lies in wait for us, lies in wait for you, in each line of the book: "Those looking for adventure will find it—in proportion to their courage."
—CARLO OSSOLA

This book urges us to rediscover the freedom to create using the incredible wealth we all have available to us: that of our brain, our body, and the broad knowledge that is more easily accessible today than ever before. I suggest that vicariance—the substitution of one process for another, but leading to the same goal—is one of the fundamental mechanisms of life. Vicariance gives living organisms a powerful capacity to create, innovate, and interact with flexibility, tolerance, and generosity. Vicariance is also the key to the remarkable human faculty for creating imaginary worlds. These are the worlds that individuals invent during the course of everyday life and that allow them to transform their desires and beliefs at will. They are also the worlds that the imagination creates in novels and fiction, as well as those to which we have access thanks to modern imaging technologies and so-called virtual communities, both fascinating and disturbing, into which we can "teleport" by developing avatars of ourselves. Finally, these are the worlds we know as utopias.

Modern society is in the throes of massive changes. One extreme is pitted against another in confrontations whose outcome is uncertain. In this combat, the amazing diversity of cultures and individuals has been flattened by the need to standardize outcomes and behavior. For example, economists now represent the human being by equations, favoring financial cost over individual creativity. They have stifled pride in resourcefulness through a multitude of production constraints (barbarously referred to as "process").

This work is of a piece with my previous books, *The Brain's Sense of Movement, Emotion and Reason,* and *Simplexity.*[1] For this reason, you will find that it refers to some ideas and themes that I have already discussed. But here you will find them in a new and, I hope, stimulating light. The painters of the Renaissance generally chose familiar material for their subjects, but innovating each time; in the same way, the thoughts of adults often echo those of children. The development of the brain is, in fact, an ascending spiral, each step giving the impression of going backward, but at a more developed level. One could also reverse the spiral and say that every new treatment of a question in turn makes it possible to delve more deeply into it and to shed new light on it. In any event, that is the challenge that I hope to meet. In *Simplexity,* I sought to show how greater complexity (elegance) can lead to simpler solutions to complex problems. In the present volume, I propose that vicariance provides new solutions by substituting one solution for another to solve a given problem, or by using the solution to one problem to solve another problem. These vicarious transfers make it possible to reach the same objective.

I also refer to (and cite where possible) documents I consulted here and there, including on the Web—published papers, personal communications, and so forth. Some work is directly quoted, some paraphrased. I do not mean to suggest that these theories and findings are definitive. The mathematician Henri Poincaré said of Euclidean geometry: "It is not the truest, but it is the most convenient." Similarly, the research and ideas presented here are the most useful for illustrating and supporting the book's basic theses. Please forgive their variety, their eclecticism, and their incompleteness. If the unity of the ideas is not readily apparent, the fault is mine. Since I am a physiologist, I describe the neural basis of diverse forms of vicariance without claiming that they are basic in the ontological sense, that is, the very foundation of lived experience. The partial knowledge that we are able to gain regarding the functioning and cerebral workings of behavior must still be considered in the context of individual experience.[2] This is why I sometimes make excursions into the domains of literature, art, and even anthropology.

Introduction

Vicariance: A Many-Faceted Concept

The word "vicarious" comes from the Latin *vicarius,* which, strictly speaking, means "substitute" or "replacement." *Vicarius* itself derives from *vicis,* which means "change." The notion of replacement is an extension of the word's initial Latin meaning and its Indo-European root, *weik,* which means "to turn" or "to bend." This root crops up in many words expressing change, such as "vicissitude," but with different meanings. "Vicarious" is sometimes used to refer to social dimensions in relations with others. For example, *vicarius* denoted a sheriff in England in 1066 and the ruler of Italy around 1400. The vicar may serve in place of a priest or bishop. He can say mass and attend to administrative business. However, the extent of his spiritual powers is limited. As far as the church is concerned, matters pertaining to the religious order—for example, the ordination of a priest—come under the realm of the sacred and thus do not concern him. He is not quite the alter ego of his superior. A colleague in paleontology used to tell me that his first job, at a museum in Stockholm in 1972, was as a *vikarierande amanuens,* or "substitute conservationist." The idea of substitution also figures in "viceroy" and "vice president."

The concept of vicariance is not a familiar one, and the word, I admit, is not especially pleasing to the ear. But it is widely used, and not only within the confines of academia. Vicariance turns up in important areas of modern social life, such as the question of multiple identities. It is involved in the compensation of deficits in neurological disorders, navigation around a city, reasoning, education and learning, architecture, industrial design, diversity of opinion, tolerance, and (finally) the capacity to create and to innovate. In French, the term *vicariance* shares a common religious origin with the term *ordinateur* (computer). As the religious meaning of *ordinateur* (one who ordains priests) fell out of fashion, linguists appropriated the word to translate the English term "computer."

You might well ask what is so interesting about a concept as vague and multifaceted as vicariance.[1] Consider the following question: Among the many meanings of vicariance, is there evidence suggesting an important foundation for the relationship of "life" with the world, with people and things? Do the various meanings of the word "vicariance" have some common origin that even people who use it have failed to recognize? Could this age-old concept actually be a very modern idea that reflects the transformations humans are experiencing in every area of their personal and social lives? Could it be a fundamental principle of life? In short, is the word "vicariance" only good for Scrabble, or could it change the way we think? I previously included vicariance in a list of the essential principles of simplexity.[2] Here, my intent is to delve more deeply into the meaning and use of the concept, and to reveal the complex mental processes that it encompasses.

The main challenge for anyone interested in vicariance is the abundance of ways in which it is understood. I will explore these in more detail later. But allow me to summarize them quickly to give you a feeling for their diversity.

Vicariance has been used to describe the variety of living forms, from a simple copy—cellular differentiation (mitosis) or gene replication—to the variations generated by these copies. In biology, "alveolar vicariance" describes the mechanism by which the lungs take in air through parallel channels. These channels are mutually replaceable. A "vicarious host" is a living organism that shelters a parasite, even if only from time to time.

Paleontologists employ the term *biogeographical vicariance* to designate the diversity of animal species occasioned by continental drift. They also refer to *ecological vicariance,* that is, species diversity caused by climatic and environmental changes.

When psychologists speak of *vicarious functioning,* they mean the capacity of one function to substitute for another. For example, you can cross a river in a plane, in a boat, or under your own power, by swimming. Similarly, to maintain the body's balance, the brain can use visual, muscular, vestibular, or tactile information, or combinations of these data.

Among linguistics scholars, the term has been used to denote certain uses of some words, especially in French, such as the subordinating conjunction "that" (*que*). In a sentence with two subordinate clauses, *que* can be used to introduce the second clause in order to avoid repetition of the first conjunction, as in "*Comme il faisait beau et que je n'étais pas très occupée, j'ai fait une promenade*" ("Because it was a nice day and that [in place of repeated "because"] I wasn't very busy, I went for a walk"). Here, it is called the "vicariant 'that.'"[3]

Vicariance also turns up in the work of the ethologist Jakob von Uexküll, in association with the idea of *Umwelt,* that is, the fact that each living species constructs its world and attributes different meanings to things based on its capabilities and the surrounding environment. The same object may be put to very different uses, each of which gives it a different meaning. For example, a knife is used for cutting, but it can also serve as a screwdriver. A flower stem provides a ladder for an ant, food for a cow, and so forth. I call this *vicarious usage*, and I will return to it later.

In these pages, I will explore the idea that vicariance is involved in the sharing of emotions, as in empathy. I will look at the use of avatars in virtual worlds—virtual vicars! One might even say that substituting money for objects or services also qualifies as a vicarious process.

Vicariance is a simplex principle, a work-around created by diversity. A fine example, described by the neuroscientist Jean-Pierre Changeux, is the proliferation of neurons during ontogenesis. The apparent confusion resulting from this proliferation enables specialization of neurons and associated networks. This diversity involves a fundamental property of life: survival by sidestepping reality, by escaping the rigid constraints of rules. How? Through the capacity of the human brain, by virtue of evolution, to find novel solutions to problems posed by interaction with environmental forces or with other humans, and consequently to create possible worlds. Alain Prochiantz, a biologist at the Collège de France, recently referred to the gift that evolution gave us of 900 "excess" grams of brain.[4] I suggest that vicariance is one of the many new attributes acquired as a result of this fantastic increase in brain volume. Evolution favors duplication. Duplication increases the processing power of the brain, even if it sometimes comes at the cost of catastrophic proliferation, as in the case of cancer.

But vicariance is much more than simple duplication: vicariance is an inventor. Social norms free the rabbi to devote himself to interpreting the Talmud. Perhaps the success of the concept also derives from delegation to the vicar, which frees the priest and allows him to escape from mundane tasks. To my mind, it is this model—partly implicit—of creative liberation that explains the success of vicariance. Vicariance gives a name to the workings of the brain as a *creator of worlds*. These worlds provide a stage for anticipating the future, and even for shaping it. The use of imagination makes it possible for the brain not only to simulate or to emulate, but also to speculate. It is for this reason that I include vicariance among the principles of my theory of simplexity. The first three parts of this book attempt to tease out the facets of vicarious processes in brain functioning and in our relations with others. In the fourth part, we will see how the concept can be used to modify the standard approach to learning. Finally, in the Epilogue, I return to the connection between vicariance and creativity.

This book is ambitious, and there is a potential for limitless semantic multiplication. The trick is to bound the concept without limiting it too much and thus losing its creative potential. It is also necessary to distinguish between vicariance and representation. That in turn will require understanding both the link between its uses and the source of its richness.

From the Universal to the Particular

Another hypothesis I will explore is the role of vicariance in the relationship between the universal and the particular. For example, in the area of human rights, this relationship manifests as the tension between the rights of the individual, so strongly championed by lawyer and Nobel Peace Prize laureate René Cassin, and universal laws. In linguistics, it is evident in the debate over the diversity of languages, so brilliantly championed by Claude Hagège, versus the universal rules of grammar established by Noam Chomsky.[5] In mathematics, we see it in the contrast between mathematics founded on axioms and those inspired by intuition, defended at the beginning of the twentieth century by the "intuitionist" mathematicians.[6] This gave rise to the famous rivalry between David Hilbert and Henri Poincaré. In pedagogy, it is the difficulty in combining the development of each student—as the physicist Georges Charpak managed so successfully with his project for hands-on science education (La Main à la Pâte)—and standard methods of teaching, which are often too abstract. In anthropology, it is the opposition between the "structuralist" approach and the ethnographic description of real-life experience and mythologies. In social life, it is the complex relationship between specific and collective interests, and in juridical matters it is the contrast between jurisprudence, legal exceptions, and general laws. In the world of work it is the conflict between work as it is actually done by each individual and assigned tasks.

Unfortunately, the great debate—feud?—between the particular and the universal has explicitly favored the latter and either forgotten or denied the diversity of individuals and their cultures, the flexibility of their brains, and the creativity of their imaginations. In academia, a wave of rediscovery and "naturalization" that champions action, diverse perspectives, and interpretive plurality is cutting across various disciplines. Examples include French linguist Alain Rabatel's work on the importance of "point of view" in language,[7] as well as the studies of geographers Jacques Lévy and Michel Lussault. Lussault writes of the occupation of cities by groups of activists: "Occupation undermines the edifice of democratic representation that was founded on the remoteness of representatives, their absence from everyday life, on the production of an isolated and isolating political machine that defines very limited and ephemeral relations with citizens."[8] In *L'Homme rhétorique* [*Rhetorical Man*], Emmanuelle Danblon, professor of rhetoric at the Free University of Brussels, writes, "A widespread tendency in nature is to welcome the unexpected, to adapt to it, but also to search for new solutions to recurring problems. In the naturalist vision of rhetoric, . . . it is necessary to regard human reason as the result of a stratification of natural faculties that are regularly readapted to new institutional realities."[9] From this perspective, a major, age-old conflict pits reason against emotion, determinism against uncertainty, the universality of general laws against the diversity of specific examples, in the living world just as in the legal system. Vicariance is one of the simplex mechanisms invented by evolution to resolve this conflict.

All areas of social life are currently undergoing radical change. The complexity of the modern world has stimulated a growing desire for standards. The origin of the desire for standards and norms seems obvious. It stems from the amazing invariance of natural forms, from quartz crystals to living creatures, which suggests the existence of universal laws. In addition, individuals in society saw standards as a condition of community life: there is no democracy without rules. Standards were adopted very quickly in the organization of work and technology. Even games have rules, and the constraint of notes is no impediment to music.

Today, there is a resurgence of interest in the singularity of individuals, the diversity of their behavior, and the solutions they are capable of inventing to meet life's challenges. In 2013, in Geneva, a consortium of physicians and international institutions organized a conference on "person-centered health" (and we thought medicine was already focused on the patient!).[10] The international transportation industry has pledged to put travelers back at the heart of the system. With the development of the field of neuroeconomics, economists are suddenly interested in emotions and in how individuals make decisions. Finally, eleven Italian universities, following the lead of education professor Maurizio Sibilio of the University of Salerno, recently discussed the possibility of selecting my book *Simplexity* as a guide for rethinking pedagogy and interdisciplinary didactics, with a focus on action and individual strategies—as opposed to the normative orientation of theoretical didactics. This list of examples illustrates the extent to which the individual has been neglected. Norms limit creativity and the potential of each of us to act. Action, know-how, and the body itself are coming to be seen as the wellspring of thought, and there is a move afoot to give individuals the power to affirm their particularity. Vicariance is the perfect tool for this transformation.

Even artists maintain that they sometimes need rules to transcend rules. At the imaging arts and technology department at the University of Paris 8, artists Michel Bret and Marie-Hélène Tramus programmed an artificial tightrope walker. Its movements were very stilted, and I helped them to incorporate the laws of natural movement into the program.[11] The goal wasn't to make the tightrope walker more "natural." Rather, the point was to enable the researchers to use these laws in order to break them and thereby induce emotion, or surprise. The conflict between law and diversity gives rise to aesthetic experience. Or, to put it another way, the tension between diversity and universal laws can be creative, and vicariance is one of the preferred tools of this dialectic richness.

Finally, leaving aside scientific facts, I will conclude with the marvels offered by the many shades of vicariance. On the subject of Shakespeare and the comedy of wonder, the poet Michael Edwards has written, "It pleases Shakespeare not to express himself, not to say what he thinks, but rather to explore what others are thinking, to be privy, in the very act of imagining his characters, to new ways of being, to come to know a host of possible worlds."[12]

Part I

The Vicarious Act

1

The Brain as a Problem-Solver

Vicarious Functioning

Humans have invented many ways of accomplishing certain tasks, such as crossing a river. We also have many cognitive strategies for remembering the route we take from home to work. Modern navigational aids (such as GPS) offer us the choice of viewing a sequence of steps or of seeing a map. We also have different tricks for remembering names. To give just one example, my father taught me a mnemonic for remembering the names of the Roman emperors: "CAESAUTICA GLAUNEGALO VIVESTIDO."[1] And rhetoric provides us with a range of approaches for making an argument.

One of the many meanings of "vicariance" involves the idea of substituting one mechanism or process for another to reach the same goal. The term is not new. In medicine, from the late nineteenth century onward, it is mentioned in connection with replacement of the function of a defective organ. In his *Traité du caractère,* Emmanuel Mounier wrote: "Modern psychophysiology teaches us that where specific connections exist between certain regions of the brain and certain psychic functions, vicariance enables a healthy region to take over the role of an injured one."[2]

Influenced by the theories of his cousin, Charles Darwin, Francis Galton pioneered tests and systematic methods to describe individual differences in intelligence. He also considered variation to be an important aspect of evolution and of its adaptive nature. In 1895, Alfred Binet and Victor Henri sketched out a new form of "individual" psychology that corresponds to what we know today as differential psychology.[3] They described the practical and theoretical issues involved in the study of individual differences, and they explained why "mental tests" (then beginning to be used in schools in the English-speaking world) were ill suited for this purpose. The most important and interesting differences between individuals, they maintained, are observed at the level of higher psychic processes such as intelligence, and not at the level of elementary processes (feelings), which these tests measured. Their work contained the germ of the ideas that led them to publish, ten years later, the first test for measuring intelligence. Thereafter, a schism developed between experimental psychology and comparative psychology that lasted until Lee Cronbach's efforts in 1957 to unify the field.[4] Even Marcel Proust had something to say on the subject, writing in *The Cities of the Plain,* "Rheumatism and neurasthenia are vicarious forms of neuro-arthritism. You may pass from one to the other by metastasis." He also referred to "habits that will, one day or another, return to fill the place of the vicarious, cured malady."[5]

Around 1950, a group of psychologists led by Maurice Reuchlin, Jacques Lautrey, and Théophile Ohlmann developed an approach to differential psychology that was concerned with the various perceptual and intellectual strategies an individual could use to solve a particular problem.[6] According to Reuchlin, the issue was not one of vicarious categories but rather one of vicarious processes. The concept of vicariance developed by this French school was in line with functionalism: all the behaviors that fulfill the same function can be considered as equivalent and are thus mutually substitutable. Each individual has recourse to several vicarious processes for adapting to any particular situation. Some of these processes are less easily evocable than others; that is, they exist in a state of potentiality in the brain and its networks but must be triggered. The hierarchy of evocability differs from one individual to another as a result of genetic makeup, previous experience, or the interaction between these two categories. Interestingly, Reuchlin also discussed the fact that these individual differences are linked to general laws, a problem that I will discuss later in this book as being crucial for creativity. Vicariance entails a cognitive cost that is determined by the probability of success of the process (less costly when the probability is high), the time it takes to implement and use the process (less costly when time is short), and how automatic the process is (less costly if it is automatic). Finally, if a process has been used and proved satisfactory (one way or the other), it will be used again. This is an example of positive reinforcement; negative reinforcement (the inverse) also occurs.

The heuristic value of this theory has been shown in the area of spatial perception, for example, the perception of the vertical that is provided by the vestibular system, by vision, or by a combination of various senses. Another example involves the shift in cognitive strategy employed by an infant attempting to determine the number of objects lined up on a table. Very young babies assess the number based on the length of the series (confusing length with number). Later in their cognitive development, they suppress this strategy and count the number independent of space. Other examples can be found in the areas of memory, classification, and written language.

This approach was discredited by research on the general laws that govern the brain. The advent of statistical methods meant defining standard behavior. However, the differential approach has recently returned to favor. This is due in part to brain imaging, which has shown that different people use different configurations of neuronal networks to solve the same problem and, in addition, that depending on context or intention, the same person may use different networks (that is, different cognitive strategies) to accomplish a given task.

The Differential Approach: The Return to the Individual

As I mentioned in the Introduction, we are witnessing a dramatic criticism of the formalist approach that is currently dominant in every area of human activity. Since the appearance of my book *Emotion and Reason,* I have been invited to conferences in a range of milieus, including industry, psychiatry, education, transportation, and banks and insurance companies. I hear the same criticism everywhere: we are neglecting individual experience, specific women and men along with their differences and their creative potential. Returning to a study of inter-individual and intra-individual differences is thus fundamental. The desire to identify an "average person" can be situated in the more general context of the dominance (for more than a century) of a mindset—particularly in France—that favors abstract thought, the absolute, and centralized power. As Jean-Paul Sartre wrote, "*Means,* as Engels and statisticians conceive of them, suppress the author, but by the same stroke they suppress the work and its 'humanity.'"[7]

I assert that the distrust of the British (and by extension the Americans) toward the dominance of the normative, rational, abstract thought that triumphed in France, along with their confidence in the ingenuity of young people and their pragmatism (inspired by William James), are in part responsible for their commanding role in science. This mode does not inhibit the development of theories, but the theories are always anchored in reality. Except, of course, in mathematics, where the abstract thought so dear to the French and Russians has fueled the success of the Fields Medal. English and French express action differently. In English, you say "swim across the river": the action steers the abstract goal. In French, we say "traverser la rivière à la nage" (cross the river by swimming): the objective, the idea, precedes the action. Hence the power of theory, but also the risk of cutting oneself off from reality!

To be clear: I do not deny the enormous utility of abstract thought. It is beyond question. The psychologist Johnson Laird, for example, was a firm believer in the amazing effectiveness of diagrams. The goal here is not to pit empirical thought against theoretical thought, but to go beyond that duality to demonstrate the richness of possible solutions.

Vicarious Usage: Von Uexküll's Worlds

The ethologist Jakob von Uexküll (1864–1944) suggested a radical inversion of the traditional meaning of vicariance.[8] According to him, the perceptual world (*Umwelt*) is narrowly circumscribed. For example, all that a tick knows of the world is butyric acid and heat. When, from its position on a tree branch, it perceives an object that possesses these two properties (which signal animal flesh), the tick drops onto the skin it has detected and pierces it, then sucks the animal's blood, lays eggs, and dies. That is a very extreme and limited form of life!

Von Uexküll defined what I call *vicarious usage*. He means by this the capacity of living organisms to exploit their immediate surroundings in very different ways depending on their own goals and their *Umwelt*. The best-known example is that of a flower stem, which is transformed by the five contexts of a young girl, an ant, a larva, a cicada, and a cow. It has a meaning for each because they use it, respectively, as an ornament, a means of access, a construction material, or food. The same object is thus applied to a variety of processes, intentions, and usages in the full sense of the word, as encountered today in the field of design, for example. The object is not duplicated; there is no substitution or replacement. It is perceived and applied vicariously to different ends.

Humans and other large animals are different from ticks. But they, too, are constrained by a fairly limited repertoire of perceptual responses to their *Umwelt*. This term is sometimes translated as "milieu"; but the translation is incorrect because it signifies how the animal uses its surrounding environment, rather than the opposite. Indeed, for von Uexküll, living organisms do not consist of a collection of components that combine to make a whole. They are constructed according to a goal, a bona fide purpose that determines the parts in a centrifugal rather than a centripetal way. Action is what gives artifacts ("facts of art," that is, objects, tools, and machines produced by humans) their meaning. Von Uexküll cites the example of a young man who had never seen a ladder before and perceived it as a collection of bars and holes. Meaning, von Uexküll wrote, is always part and parcel of the action one intends to execute with the object, which simplifies perception and, especially, organizes the world. Von Uexküll distinguishes two types of perceptual processes. The first is linked to experience, which etches in memory the signals perceived during a sequence of actions in the real world (for example, memorizing a route through a city) and is thus mediated by the senses. The second, in contrast, is innate, in that how the world appears is determined almost "magically," as in the case of the flower stem. These signals are linked according to an "inborn melody." In the context of their *Umwelt*, animals can create imaginary worlds, which von Uexküll called "magical." These worlds allow predictable behavior. Paradoxically, this selection of perceptions and behaviors by each species, though it is a constraint, also enables a new freedom: the possibility exists to substitute not one person for another, as in the case of the vicar, but one application or function of an object for another. This is a particular form of vicarious functioning.

For von Uexküll, the same object may therefore be perceived in several different ways. Action determines meaning in the repertoire of our *Umwelt*.[9] He also emphasizes that living organisms accomplish tasks in a variety of ways. This is precisely the property of vicarious functioning that I described earlier. The psychologist Henri Piéron has also discussed the diverse worlds created by our brain.[10]

Vicarious usage thus attributes multiple functions to the same object. This idea can be applied to the function of objects that we use ourselves. In the same way as in the examples of the ladder and the flower stem, a knife can function as a screwdriver, a chair as a stepping stool, a hat as a collection basket, and the wall of the living room as a place to hang pictures. In everyday life, we also attribute different roles to others. If I ask a shopkeeper for information when I am lost, she transforms from a salesperson into a guide. A physician may also be a mother or a father of a family. The fact that one object can be used in various ways was also touched on by the Russian psychologist Lev Vygotsky (1896–1934). A hand can be used for grasping, but also for signaling.

The concept of vicariance thus designates two different, but very creative processes. The first, which I call *vicarious functioning,* indicates that the same task can be performed in several ways. The second, *vicarious usage,* indicates that the same object or part of the body, or the same person, can be perceived as fulfilling different roles based on our intentions and our *Umwelt.* Note that vicarious functioning and vicarious usage may be present simultaneously, for example, if I am drawing on a piece of wood with a knife.

Biogeographical Vicariance: Diversity or Innovation?

Everybody knows that frogs and butterflies in Africa, Europe, and the Americas are similar but not identical. This splendid diversity fascinated the great mathematician Laurent Schwartz in the case of butterflies, though his research primarily involved abstract universal concepts. I myself was fascinated by the fact that he carried his butterfly net to every conference he went to. He assembled an extraordinary collection.

Evolutionary science employs the concept of vicariance to characterize the diversity of animal species. But the field, too, is diversifying. Paleontology, for example, employs the term "biogeographical vicariance" to describe the diversity of living creatures and their adaptation to the planet's environmental changes.

These changes are due in part to continental drift. Scientists have conceded that Alfred Wegener was right: from the Carboniferous to the Jurassic—roughly 350 to 200 million years ago—most of the continents were unified in a single continental mass called Pangaea.[11] The fragmentation of this protocontinent, which continues today, occurred following the appearance of many species that later became separated by geographic barriers. This separation, brought about by plate tectonics, resulted in variations of form and behavior in species with common ancestors. A very good example of vicariance is that of the Dipnoi (lungfish) of Africa and South America, which have a common ancestor but evolved separately following the opening of the South Atlantic.[12]

Vicariance is a key concept in the study of evolution. However, the elucidation of its mechanisms has given rise to many controversies. The in-depth analysis of the forms displayed by different groups of species (specialists refer to these groups as "clades," and accordingly to cladistic analysis) suggests that a mechanism of dispersal within continents, or even across oceans, might explain the similarities. This process is known as *dispersal vicariance*.[13] Paleontologist Philippe Janvier writes: "Seeds carried by the wind or floating in the sea, insects borne away by storms, natural rafts spewed out by large rivers have for a long time (and even today) allowed species to cross the nascent Atlantic. In this way, perhaps transported by natural rafts, primates and African rodents were able to colonize South America around forty million years ago, when the South Atlantic was already open. Sorting out which aspects of biotic diversity are related to continental drift and which were generated by random dispersal is difficult to do, but it is possible: This is the aim of 'the biogeography of vicariance.' "

Ecological Vicariance

Another use of the concept of vicariance for animal species applies to a species that, in a given region, replaces another species that occupies the same ecological niche and inhabits a similar environment. A classic example of this *ecological vicariance* is the African dwarf hippopotamus (an artiodactyl) and the capybara of South America (a large rodent related to the guinea pig). Both are about the same size and are similarly adapted to living in swamps. In geographically distant but ecologically analogous areas, one sometimes finds species or broader groups that are taxonomically close.

In short, the term *biogeographical vicariance* refers to common origin, and *ecological vicariance* refers to similarity within diversity.[14] In other words, and this is a major theory of this book, the tension between universal properties and diversity, having to do with continental drift and with numerous climatic and environmental factors, has had a powerful creative effect. The concept of vicariance implicitly embodies this creative power. One might well object that biologists and paleontologists chose the word "vicariance" for want of a better one. However, I may answer that anyone who believes that a concept has only one meaning will fail to appreciate the point of this book: the capacity to create, generated by diversity and biological or semantic variations. If the word "flame" were incapable of simultaneously designating the fire burning in the hearth and the passion of love; if humans were deprived of metaphors; if rivers could not slake our thirst, carry our barges, harbor newborn salmon, feed the turbines of hydroelectric plants, inspire the crazed meandering of Rimbaud's drunken boat, and lead Eurydice away from the torments of Hell, what a dull world we would live in!

2

Perceiving and Acting

Among all of these different forms of vicariance, it is important to remember, for our purposes, the essential elements: replacement and substitution. But already implicit in the concept is an aspect that enriches it considerably—that of the creative function of diversity.[1] The separation of continents was a factor in the diversification of species. In conjunction with genetic and epigenetic mutations and adaptations to the environment, it created an extraordinary diversity of forms and of adaptive solutions.

Although not always explicit, the notion of vicariance is inevitably related to the notion of action. The vicar replaces the priest at worship services. Biogeographical and ecological vicariance, both sources of biodiversity, presuppose adaptations related to the planet's drifting land masses, to new environments, to competition for survival, and so forth. *Vicarious usage* concerns different actions involving the same object, and *vicarious functioning* refers to substitution of one function for another (there are several ways of doing the same thing). In previous publications, I have emphasized the profound significance of this concept of action.[2] It integrates perception, movement, and various elements of intentionality (including

motivation, emotion, and the interplay between the past and prediction of the future)—whence the title of this chapter. My aim is to
describe the component processes of vicariance, from the most
individual ones to those concerning cultural and social life. The following sections present examples from modern neuroscience.

Vicarious Perception and Movement

At night, when you are struggling to move around your room in the
dark, the brain relies not on vision but instead on memory of places,
motor habits, and the sense of touch. It is a wonderful example of a
vicarious process. One sensory experience can replace another, or
a combination of others, in order to formulate a perception that is
relevant to the goal or the context. I call this process *vicarious perception.* I use the term "perception" rather than "sensation" because
vicariance presupposes that the function of the defective sensory
process is replaced; thus a new perceptual process is put into play.

Sensory Substitution

There are several categories of sensory substitution. One category
is natural physiological substitution. It occurs spontaneously, for
example, when we are moving around in the dark at night, as I mentioned above. We employ other sensory clues in the place of vision,
as blind people do. Another example is what happens when judging
the properties of an object, say, a seashell on the beach. We can do
this either by using vision alone with the help of different visual
analyzers (texture, shape, color), or by using a combination of sensory receptors (vision and touch). These two modes of analysis are
obviously never completely equivalent. With certain combinations,
we can either gain or lose information.[3]

Living organisms anticipated the possibility of sensory deficit by
duplicating the coding of variables relevant to perception, thus

creating functional redundancy. The psychologist Théophile Ohlmann has studied vicarious perceptual processes in depth. He notes, for example, that texture can be coded in three ways: through vision, touch, and hearing. A passive, relatively rapid displacement of the head is detected at least three times: at the vestibular level by the semicircular canals, at the visual level by the optic flow appearing on the retina, and finally by the proprioceptors in the neck. Patients presenting with lesions of the vestibular sensors, which measure movements of the head, use visual information or muscular proprioception to compensate for the deficit.

This redundancy poses a problem of compatibility, of "coherence" between different modes of encoding, but it also has advantages, such as robustness and the resolution of ambiguity in the information given by the sensors in an uncertain world. Vicarious sensory perception is thus not a simple duplication. It can also be a means of verifying the quality of the sensors and eventually of selecting the combination best adapted to the action under way. For the neurophysiologist Jacques Paillard, far from being a disadvantage, vicariance represents a genuine source of inter-individual diversity by virtue of its combinatory potential.[4] It constitutes a "reservoir of capacities for adaptation" through the diversity of its duplication of functions or of solutions. It is also an example of the importance of vicariance for creativity.

The Vicarious Gesture

Our actions are expressed by gestures. I have discussed the richness of gesture in my book *Simplexity*. Here, I would like to emphasize the incredible multiplicity of functions and expressive capabilities of gesture. Think about grasping an object. The neuropediatrician Giovanni Cioni has described the steps in a child's acquisition of different strategies for grasping, strategies that gradually replace one another.[5] You might call this *vicarious movement*. First, the infant is able to use a small repertoire of grasping patterns selected on the basis of the affordance (the "graspability") of objects. For example, it can manipulate its bottle with great dexterity. The infant employs these patterns in a random fashion. This repertoire gradually increases. Palm grasp appears around 24 weeks, when the hand surrounds the object via flexion of all the fingers. Around 28 weeks, the infant grasps by pressing the object with its fingers, ensuring stability with its thumb. A differentiation occurs within the hand: around 32 to 36 weeks, the object is grasped with the fingers alone, without using the palm; and, finally, the infant is able to grasp the object between its fingers and thumb and, around 40 weeks, further refine the use of its fingers. Cooperation between the two hands is established later.

An interesting form of vicariance is related to the fundamental symmetry of living organisms. This subject is worth a lengthy discussion, but an example borrowed from manual grasping will suffice here. Humans have two hands controlled by the "two" brains: the left hand is controlled by the right brain, and the right hand is controlled by the left brain. The functional lateralization of the two hands and the two brains is well known. However, it is possible to replace the use of the primary hand, usually the right, with the other hand. I call this *gestural vicariance;* each hand is thus potentially the vicarious substitute for the other.

Rehabilitative specialists like Moshé Feldenkrais have used this vicariance between the left side and the right side in motor retraining of patients with movement disorders of one side or the other.[6] In his book *Body Awareness as Healing Therapy: The Case of Nora,* Feldenkrais reports that he was asked: "You stressed the difference between Nora's recovering her writing ability and re-creating the same ability. How can you re-create an ability?" To which he answered: "An ability results from a structure being trained until a skilled functioning is obtained. The ability is erratic and rudimentary in the beginning. In due time and with sufficient practice, it begins to operate more and more smoothly. During that time a large number of cells in the nervous system have been involved. When an ability cannot be performed as before, only some of the cells which were essential to the skillful performance have ceased to function. But there remain a multitude of cells that were auxiliary to the performance. In some cases it is possible to lead these auxiliary cells to play the role of the principal cells, though usually in a different manner. *This involves re-creating something, not restoring it to what it was before.*"[7] The examples provided above on sensory motor substitution mechanisms deal more with simple redundancy, whereas in the case of the hand, the substitution entails a genuine central reorganization.

A gesture is an action, but it is also a symbol. It is clear that the evolution from action gesture to signifying gesture—a sign or symbol—endows it with a new, remarkable property of flexibility, of *symbolic vicariance*. As Frans de Waal has shown in the case of primates, gesture can threaten, but also reconcile.[8] It is an essential tool for perceiving and interpreting the emotions of others.[9] But here, too, vicariance is a source of creativity, since symbolic gesture is also used in religion (as in the static gesture repertoire of representations of the Buddha), theater, regulation of traffic by police officers, and by photographers, painters, and sculptors to various creative ends.[10] The hand may thus serve to grasp, but also to signal. Renaissance painting, which often erased facial expressions of emotion, relied heavily on gesture. One of the most beautiful examples is the annunciation of the Virgin. Despite an impassive expression, one detects submission, surprise, astonishment, and even denial. We humans constantly use our hands and arms to express intentions and emotions. We are dealing here with a very general and basic form of vicariance. In Chapter 9 I touch on the application of these ideas to learning by Lev Vygotsky and to the mental gestures theory of Antoine de La Garanderie.

Cyborgs

A second category of vicarious processes is substitution by implanting on the body, or even inside the body or the brain, various sensors and sensory or motor prostheses. This modern form of vicariance by hybridizing natural and artificial components will become more widespread in the future and thus is worth a word here. We call a rebuilt human a "bionic" person or a *cyborg* (cybernetic organism).[11] A cyborg is a living creature (in this case, a human) equipped with artificial devices that enhance its capacities or record its vital signs (heart rate, blood pressure, and so forth). Anyone wearing a cochlear implant or even sunglasses is a cyborg! Literature and film have been creating fictional characters of this type for a long time.

An example of the success of such technologies is an artificial retina that can be implanted on the retina of visually deficient patients, developed by ophthalmologist José-Alain Sahel and his group.[12] These devices allow the patient to recover much more visual function than was expected of this stimulation, which is quite minimal compared with the normal activity of retinal neurons. By the same token, direct stimulation of the cortex did not produce the desired effect.

Another good example is cochlear implants, which electrically stimulate the inner ear (cochlea). These implants work very well: patients succeed in reconstructing what another person has said, even though the electrical stimulation by the implant only partially reproduces the neuronal activity corresponding to a normal ear. The brain, thus, can infer a complex sound composition from sparse data. It can even use visual information for hearing.[13] Deaf people, for example, read the lips of the person talking to them. In these people, the areas of the brain normally activated by auditory information are activated by visual information from the movement of the lips of their interlocutor. The auditory areas are said to be "colonized" by the information produced by the visual system. People who have been wearing a cochlear implant for more than a year show both improvement in hearing and a sharp reduction in colonization of the cortical areas for voice through the visual processing of lip-reading. At the same time, the temporofrontal circuit linking the auditory areas to the language area (Broca's area) is reactivated.

Stimulation of the vestibular nerve may also result in compensation for deficits caused by damage to the vestibular sensors, the semicircular canals, and the otoliths. We recently collaborated in a European project, coordinated by Silvestro Mircera in Paolo Dario's laboratory in Pisa, to develop a vestibular implant in patients with peripheral damage to the vestibular receptors.[14] In this context, we proposed a completely new theory regarding how the otoliths of the vestibular system function. Several implantations have already been carried out in Geneva and Maastricht. Here again, vicariance implies extremely sophisticated cognitive development because patients must use information that is very different from that normally supplied by the biological sensors to perform the same function. They must manage to recover not only the functions of gaze control and balance but also the cognitive functions associated with the vestibular system (spatial orientation, construction of self, and so forth). Several groups are currently working on these projects.[15]

Seeing and Perceiving with the Skin and Muscles

Blind people can see with their skin. In the 1960s, the physiologist Paul Bach-y-Rita devised a sensory substitution apparatus for the blind.[16] A video image of a scene activated a collection of vibrators on a plate placed in contact with the back of the blind subject. Bach-y-Rita's remarkable discovery was to show that when the skin of the back was stimulated, the brain could perceive (without seeing) objects placed in front of it. For example, if a balloon was thrown toward a person's face while her back was stimulated, she would perceive the balloon as approaching toward her from the front!

Nearly forty years ago, I myself had the privilege of welcoming into my team one of Bach-y-Rita's collaborators, Bill Gerrey, who was blind from birth. He had worked on a Hewlett-Packard computer assembly line with the vibration stimulator, which enabled him to assemble the electronic components. Thus, this sensory substitution is sufficient to enable a blind person to manipulate small objects that normally would require very acute vision.

These technologies are also used to allow blind people to read through stimulation of the fingertips. What is really interesting here is that with tactile stimulation, blind people acquire many of the properties of vision. There is a genuine substitution of the function of seeing. This is due in part to the visual cortex being able to directly access tactile information, but also to mechanisms (still not well understood) of neuronal plasticity.[17] Subsequently, with the evolution of the technology, Bach-y-Rita has also used tactile electrical stimulation of the tongue to replace vestibular receptors and to enable patients to recover their balance.[18]

Natural active touch can also be replaced by vibrations of the skin. The brain needs only some very simple data to reconstitute all the information, or—and this is the crucial point—to infer a reality based on partial information. For example, recent experiments show the amazing capacity of the brain to reconstruct geometric shapes based on partial local electrical stimulation of the muscles.[19] A subject was first asked to draw a triangle with a pen. Using electrodes, researchers recorded the activity of the sensory nerves of the muscles involved in the control of hand movement while the subject was drawing. Next, vibrators were placed on the muscles of the thumb that reproduced through stimulation the sequence of firings of the neuromuscular bundles (the sensory receptors that measure muscle length) during the natural motion of drawing. This artificial stimulation induces the same perception of geometric shapes that drawing would have done. The brain therefore infers the shape from the movement, despite the fragmentary information provided by the vibrators.

Finally, in the blind, the area of the brain that specializes in shaping letters can be activated by acoustic stimulation that evokes an "acoustic landscape" (or soundscape) that represents the letters.[20] These are examples of vicarious sensory substitution.

Thus, modern technology makes it possible to compensate for sensory deficits. Up to this point, we have described examples in which vicariance has been limited to substitution. However, the technology that transforms humans into cyborgs opens new possibilities. A current example is "augmented reality." The software company Dassault Systèmes has just developed a contact lens onto which information about navigation and other functions can be projected. At the cost of a detour (here, the technology), it is possible to increase the effectiveness of a human brain process. Life is using technologies outside the brain to enhance its effectiveness.

Brain-Machine Interfaces

A third category of sensory substitutes is developing rapidly thanks to the fascinating progress of technology. Brain-machine interfaces will allow the implantation of electrodes in the brains of paralyzed people in order to restore their perception or motor function, under the control of robots. Two major approaches are converging, or complementing one another. One is based on recording electric potential at the surface of the brain through electroencephalography; the other consists of implanting electrodes inside the brain. These methods call into play patients' very personality. I had the chance to collaborate with the teams of Philippe Kahane in Grenoble and Jean-Philippe Lachaux in Lyon, who introduced a method of patient control of brain activity in people with epilepsy who were equipped with deep electrodes. They called this neurofeedback technique Brain TV.[21] Patients visualize activity in a specific area of their brain on a screen and modify it by thought. They are able to change the brain region by pressing a command key.

Many teams have succeeded in commanding a robot by means of brain activity. In this case, the robot becomes an extension of the patient's body, in the same way that a pencil becomes the extension of a finger. What is strange is that for the observer, as for the patient, the behavior of the robot appears to be animated by thoughts and intentions similar to those of the patient. For the moment, these robots are controlled by the brain. But in the future, they will acquire a degree of autonomy. Such a robot would be a "vicarious partner." It is beyond the scope of this book to review this very active field of research. I will return briefly to the subject later, in the context of avatars in virtual worlds.

Two Systems for Calculation and Reading

In the preceding sections, we discussed artifacts—sensors—that enable the brain to grasp reality in several different ways. But the brain possesses other mechanisms that enable humans to approach the sensory world via multiple pathways. Consider the example of reading a text like the one on this page. We have to grasp not just the letters, but also the meaning.

Now, perform the following calculation in your head: $18 + 2 + 1939$. Stanislas Dehaene discovered that there are two distinct calculation systems.[22] One allows an exact representation of the number and the calculation using symbolic notation. The other, which is based on the analog representation of numbers, is the seat of approximative calculation. There also seem to be two systems for reading numbers: the first organizes a series of digits in a sequence of words based on rules linked to language; the second, more semantic, concerns only familiar terms that have a specific lexical input.[23] Reading numbers is thus similar to reading words, even though they are processed by different neuronal networks. I regard as *vicarious* the fact that the brain can adopt several strategies to execute the same process: approximate, rapid recognition of numbers is advantageous for action. These two strategies thus exhibit different benefits. Here vicariance is using modularity, a simplex principle.

We can find other examples of several routes for analyzing the world, such as mechanisms of emotion in the brain. When we see a snake in the forest, the amygdala analyzes only the global impression of the snake and very rapidly triggers a defensive reaction of fright. Another, slower process that occurs in a loop involving the cerebral cortex analyzes the snake in more detail and may conclude that it is harmless. Recently, psychologist and Nobel laureate Daniel Kahneman generalized this idea in his book *Thinking, Fast and Slow,* in which he describes the two cognitive systems involved in complex decisions.[24]

The dual system of calculation discovered by Dehaene led to a model of sensory substitution. Researchers Ophélia Deroy and Malike Auvray write: "Just as in reading, the use of sensory substitution devices requires training and results in the progressive automatic decoding of meaningful information. More specifically, . . . the information which was previously available and structured through one sensory modality is artificially made available through another sensory modality thanks to a code purposively designed to preserve the relevant structure and dimensions. The aim is therefore to ensure recognition of the same objects across a change of medium."[25] What is at issue here is not replacement sensu stricto. But the dual system may allow at least partial substitution. It is vicarious. It offers a choice of cognitive pathways.

A last example concerns reading sentences. My colleagues and I have shown that several processes come into play in reading depending on whether the primary interest is the meaning of the words or other elements of the text.[26] Philippe Kahane's group recorded intracranial electroencephalographic activity in epileptic patients. The subjects were asked to read a text that was either semantic, or contained "nonwords" having the same letters, or consisted of strings of alphanumeric signs (a purely visual task). The goal was to induce an activity that would be, respectively, semantic, phonological, or purely visual. The analysis of oscillations in the gamma frequency band (40 to 150 Hz), carried out by Jean-Philippe Lachaux in Lyon, showed the existence of specific networks for each task. Although this experiment does not call vicariance into play in the sense discussed here, it does show the remarkable capacity of the brain to modify these networks based on the task at hand.

The Vicariance Toolbox: Internal Models, Simulation, and Emulation

When engineers were seeking for a way to train airplane pilots, they invented simulators. Simulating turbulence required including several critical values in the computer programs responsible for the movements of the plane: the dynamic properties of the aircraft, environmental properties, gravity, wind, variations in air pressure, and so forth. Simulators thus have "internal models" for representing these values. These internal models constitute a virtual airplane, its "double." The same goes for automobile simulators, as well as for the most advanced robots. A pilot can assign a simulator low-level tasks (for example, stabilizing the plane), which enables him to concentrate on more complex tasks that require manual piloting based on the environment or instructions from the control tower. This virtual double of the airplane is a vicar, capable of acting as an automatic pilot. This is a vicarious process.

Evolution developed the same type of mechanism by creating "internal models" in our brain. These models are neuronal networks that can simulate the properties of the systems that they represent. They are found at all levels of brain functioning, from the spinal cord all the way up to the structures most deeply involved in high-level cognitive functions. For instance, researchers hypothesize that the cerebellum possesses internal models of the dynamics of the limbs. These models enable anticipatory simulation of the path of a movement before its execution. Other regions of the brain, like the parieto-insular cortex, participate in constructing internal models of the laws of the physical world, like gravity. My colleagues and I have shown that they are involved in the capacity to anticipate the action of objects on the body, such as catching a ball.[27] The new hypothesis here is that the internal models of the properties of the body, and the world, are now believed to be essential for enabling the substitution of one strategy for another.

I analyzed the functions of these models in *Simplexity*. I also showed there that the same laws of movement apply to the hand and to the generation of locomotor trajectories.[28] This property, which physiologists refer to as the "principle of motor equivalence," is an excellent example of vicarious functioning. Maurice Merleau-Ponty demonstrated his understanding of this point when he wrote that objects have an internal equivalent within us; they evoke the "carnal form of their presence."[29] This marvelous choice of words suggests that during development, driven by genetics or epigenetics, genuine internal models of the properties of objects, including an individual's physical body, form within the networks of the brain and spinal cord.[30] These internal models are not only useful for anticipating the effects of movement and for adjusting to the constraints of the task or the surroundings. They have the additional characteristic of allowing simulation of the act so that it is possible to find other strategies or to perform a different movement to achieve the same goal (vicarious functioning), or even to invent new actions that are outside the repertoire of normal functioning. In the primary definition of vicariance, the vicar replaces the priest or the archbishop; but our internal models are doubles of the world and of ourselves. The result is a new form of vicariance that might be termed "vicariance through model-making." In a certain sense, this is vicariance by constructing a "neuronal virtual" replacement. In Chapter 3, I will come back to modern forms of virtual vicariance through the use of avatars.

Internal models of the body and the properties of the physical world can be modified by learning.[31] When the models receive commands for movement, they process them as though they were parts of the body or were responding to the laws of physics. Consequently, the models make it possible to simulate the body's properties in its surroundings during action. In humans, the neocerebellum, which appeared late in evolution, may also contain internal models of objects in the world. An experiment serves to test this hypothesis. When you learn to use a new tool (for example, a computer mouse), two sorts of activities are recorded in the brain. The first is distributed throughout the cerebellum and is proportional to the motor errors we make during learning. The other is limited to the posterior fissure and is maintained after learning is complete; it corresponds to the construction of an internal model of a tool and consequently modifies the laws of functioning.

Artists have expressed in their own way some of these ideas. "Nature is on the inside," wrote the painter Paul Cézanne. "Quality, light, color, depth, which are there before us, are there only because they awaken an echo in our bodies and because the body welcomes them." Heinrich Wölfflin, a Swiss art historian, concurred: "Physical forms possess a character only because we ourselves possess a body. If we were purely visual beings, we would always be denied an aesthetic judgment of the physical world. But as human beings with a body that teaches us the nature of gravity, contraction, strength, and so on, we gather the experience that enables us to identify with the conditions of other forms."[32]

Internal models are thus powerful tools that allow the brain to avoid too rigid a dialog with reality. When we dream, for example, the activity of the centers and networks of the brain that contain the internal models of the body combine with memories and mental images. These combinations, which have given rise to many interpretations, have been for me the cause of many nightmares and imaginings that seem only too real. Internal models are a building block in the creation of new worlds. Thanks to them, dreaming is action. Von Uexküll would say that dreams unfurl in an "actant" space.

Body Schema: A Virtual Body?

To recap, we carry within our brains a double that neurologists refer to as a "body schema" or "body image." Recent work has elucidated part of the neural basis of this double.[33] It allows us to dream of surreal experiences that seem astonishingly real. If we want to understand the capacity of the human brain to interact in a vicarious way with other humans, or with humanoid robots, we must, in my opinion, understand the properties of this body schema. At issue is the representation of one's own "personal body" (*corps propre*), as in the case of phantom limbs.[34]

The expression "body schema" suggests a global synthesis of the body, whose importance has been stressed by neurologists. It is a genuine "virtual body." This concept was used by the child psychologist Colwyn Trevarthen. He suggested that a "virtual other" is present at birth in babies, and that this double awaits interaction with others. This internal simulation of the personal body is flexible and can be modified by many factors, the body schema being adaptable to the needs of the task. One must avoid falling into the trap of believing that the double is predetermined, as might be suggested by the idea that it is a simple representation of the physical body. Modern neuroscience shows that it is a process generated by action. I have discussed its flexibility elsewhere.[35]

The "rubber hand experiment" is a wonderful example of the ability of our brain to build this virtual body and even project it upon the world. The experiment consists of seating a person at a table with both arms in front. The left hand is hidden by a screen and replaced by a rubber hand. The demonstrator simultaneously strokes the hidden hand and the rubber hand. After a while, the subject has the impression that the rubber hand is their own hand. They have transferred the "ownness" of their hand to the rubber hand. More remarkable is that stimulating the rubber hand in a way that would have been painful to the real hand (for example, pricking it with a pin) also induces an impression of pain in the rubber hand "as if" it were the real one. This perception of pain is quantified by measuring the electrical resistance of the skin of the real hand, an indicator of emotional response.

Brain imaging shows that the ventral premotor cortex, the cerebellum, and the intraparietal cortex are activated by the illusion of the rubber hand, and that the intensity of the activity changes with the intensity of the illusion. With Isadora Olive I used a new experiment to study the illusion of displacement of the body schema: the "virtual piston." One of the goals of this experiment was to discover whether a virtual displacement of a part of the body—in this case, the hand—causes a rotation of the complete body schema.[36] The subject (wearing a virtual reality helmet) was told to press on a mechanical piston. A camera located behind the subject's back projected to the helmet the image of the hand activating the piston. But the projected image was shifted in the horizontal plane with respect to the real hand. A change in the body schema of a limb thus affects the entire body schema, at least in a healthy subject.

In fact, many experiments have demonstrated modifications of the neural schema of the personal body through training by using tools, even in adults and the elderly.[37] This neural schema is also modifiable under the influence of factors linked to dopamine, and the changes are all the more marked when the stimuli have a particularly strong behavioral meaning.[38]

The body schema is therefore eminently modifiable, but more important, it can in turn influence perception and action. The philosopher Jean-Luc Petit sounds a cautionary note. We must, he says, break "the magic circle of representation, which keeps neuroscience

chained to the paradigm of the mechanical brain and the intellectu-
alized body. . . . The personal body is no more the representation of
the physical body than the functional body is the representation
of the anatomical body. The true relation is the inverse; first and
foremost is the personal body, the subjective form of lived experience or the functional configuration of a living organism." Which
comes first, the physical or the subjective, is not a question that we
shall solve in this book!

The Brain as Simulator and Emulator: A Vicarious Double?

The brain is not just a simulator using internal models and a virtual
body, as in flight or driving simulators, which reproduce reality. It is
an *emulator of reality*. This concept is crucial for our central theme,
which is that vicariance creates scenarios within possible worlds.

The idea of a "phenomenal world" that is distinct from the real
world, in which our perceptions and thoughts are situated, has
been put forth by many psychologists. With respect to perception,
Albert Michotte wrote: "The role of sensory input is not, as was
long believed, to give rise to 'sensations' that can be combined, related,
or even modified by certain psychological processes predominantly
under the influence of experience (as proposed by empiricists following Hume). Its role, on the contrary, seems to be confined to the
triggering of endogenous constructive processes that obey their
own laws of organization, are largely autonomous and independent of experience, and lead directly to the construction of our phenomenal world."[39]

In the case of action, I described above the importance of internal
models of the body and the physical world and their role in anticipation and simulation of action. The psychologists Rick Grush and
Stephen Kosslyn recently proposed an "emulation theory of representation."[40] It is based on the idea that the brain contains
information-processing modules analogous to those used by robot-

icists (internal models, anticipatory feedforward control, Kalman filters, and so forth) to create, or rather to emulate, a mental environment, a world where perception and action are simulated. Mental imagery plays a specific role: it activates a process of emulation in the visuomotor networks.

Finally, modern neuroscience views the human brain as an organ that functions autonomously, an organ that is almost closed, or is enclosed in worlds that it creates itself and that can be disconnected from reality, even in healthy people. The neurophysiologist Rodolfo Llinás believes that neuronal loops linking the basal ganglia, the thalamus, and the cerebral cortex play a major role in emulation. He hypothesizes that the brain is a closed, "self-activating system whose organization is geared toward the generation of intrinsic images. . . . Given the nature of the thalamo-cortical system, sensory input from the external world only gains significance via the preexisting functional disposition of the brain at that moment, its internal context. It follows that such a self-activating system is capable of emulating reality even in the absence of input . . . , as occurs in dream states or daydreaming."[41]

Hallucinations are also creations of the brain that use the memory of lived experience (as opposed to illusions, which are solutions developed by the brain in response to ambiguities in the perceived world). They result from activation of traces of memory involved in bidirectional connections between the temporal neocortex (where the trace is stored) and the centrencephalic system (this term, a bit dated, designates the brain structures located deep within, such as the basal ganglia).

Vicarious functioning and vicarious usage thus do not refer simply to combinations of solutions that use interactions from the real world. They may use the brain's properties of emulation, via a deviation through virtual worlds.

In pursuing the ideas described above, I would like to propose an approach that might appear a bit speculative, or that might appear to extend the concept of vicariance a bit too far beyond the strict sense of the word. That is for you to judge. I suggest that the multiple mechanisms introduced over the course of ontogenesis—multisensory integration, construction of the personal body, neural double, body schema, and body image, periods where specific functions are expressed and established (critical periods), the mirror system, mechanisms of imitation and joint attention, memory, mechanisms of anticipation and internal models, simplifying laws of movement and perception, and so forth—all lead (to say "all have the goal" would be too determinative) to the capacity (philosophers would say the "disposition") to mentally simulate intentional acts. This is now generally acknowledged. Moreover, this simulation projects scenarios and invents solutions; and to do that, it creates possible worlds. Literature is the most explicit manifestation of this remarkable capacity, which is specific to the human brain. Sartre wrote: "The problem is to recover the totalizing movement of enrichment which engenders each moment in terms of the prior moment, the impulse which starts from the lived obscurities in order to arrive at the final objectification, in short the project by which Flaubert, in order to escape from the petite bourgeoisie, will launch himself across the various *fields of possibilities*."[42]

The capacity to imagine an action without executing it, using mental simulation thanks to a double of ourselves, is a very basic form of vicariance. In creating this double, the brain enables not only simple replacement but also the discovery of new solutions for the same action. I assert that mental simulation, whose anticipatory value has been studied in depth, is a vicarious mechanism of creating worlds and possible actions. The human "internal vicar" is a creator of new solutions; it is not there simply to replace us.

Ulysses's ruses are a wonderful example of vicariance. Hero of a thousand ruses, Ulysses uses the capacity of the human brain to imagine multiple scenarios. Jean-Pierre Vernant and Marcel Detienne call this capacity *mètis:* a form of cunning intelligence that relies on a variety of possible modes of action.[43] Locked in the grotto of the Cyclops, who had devoured several of his companions, Ulysses pretends that his name is "Nobody." When the Cyclops, whose one eye Ulysses has gouged out, leaves the grotto to call his brothers, they ask him who has gouged out his eye. He responds, "Nobody." They laugh and do not come to his aid. Ulysses is free to go. He then uses a second ruse to release his companions by tying them to the bellies of sheep heading out to pasture. He employs substitute strategies and detours.

At the risk of crossing forbidden disciplinary borders, I would venture that if Ulysses can use these ruses, it is because Athena takes multiple forms to help him. The principle of vicariance, in all its meanings, is pervasive in the rich tale of the *Odyssey*. Vicariance through transformation of oneself to help others is a major theme throughout literature and mythology. For example, in Japan, the goddess Kannon Bosatsu assumes multiple physical forms to come to the aid of humans. She is the avatar of the androgynous Hindu god Avalokiteshvara, "who looks down from on high" in commiseration. Kannon is often represented with ten heads and numerous hands. Each hand holds tools that she uses to help humans. She underwent metamorphosis thirty-three times to aid humanity. In China, Kannon is Guand Yin, goddess of fertility.

Mental Imagery Suggests New Solutions

To recapitulate the discussion thus far: I began with the fact that the brain has recourse to several senses to enhance perception, and even uses artificial senses in the transformation of a human into a cyborg or a bionic human. I then attempted to show that, having constructed a unity of our personal body, we have a double of ourselves, a genuine internal vicar. We will see later the importance of this double in our interactions with others. Then I proposed that the simulating properties of the human brain enable it to create imaginary worlds—to emulate reality, as in dreams—and that this capacity creates an original vicarious mechanism thanks to which the brain can delegate to our internal double the role of creator of scenarios for anticipating the future.

At the risk of overworking the concept, and provided you are still with me, I would like to pursue the consequences of this line of thinking by focusing on challenges related to moving in the environment. Over the course of evolution, living organisms have become endowed with an extraordinary capacity of vicariance in a wide range of behaviors related to moving, navigating, and wandering, from the most instinctual to the most highly cognitive. It is trivial to say that there are several possible strategies to avoid an obstacle and several ways of scoring a goal in soccer, that humans and other animals have many methods for moving (walking, running, jumping, and so forth). Humans have also devised other ways of moving, as equestrians, drivers, pilots, or astronauts. Of course, as I showed in *Simplexity,* in living creatures these behaviors are subjected to simplifying laws, but combining these laws in different ways enables development of a repertoire of behaviors for carrying out the same action or achieving the same goal.

Take the example of ducks.[44] Ever since the attempt by Jacques de Vaucanson in 1739 to create a duck-like robot, researchers have been striving to reproduce the remarkable variety of elegant behaviors typical of these birds. Ducks are equipped with an embarrassment of locomotor processes: walking, surface swimming, diving, slow underwater gliding, aquatic flying, long-distance day or night flying (in the case of some species). The diversity of these behaviors is limited by situational constraints, that is, a combination of environmental conditions and the goals of the animal. Travel distance, for example, is a decisive factor in the choice of behavior: walking, for short distances, or flying, for long distances. Thus, paradoxically, vicariance—that is, the ability to choose—is a factor in unifying the behavior of a given population. Although different individuals in a population can choose different behaviors, they generally make the same selection.

Mental Locomotion

According to a French scout song, the "best way of walking" is to put one foot in front of the other and start all over again. Actually, humans possess a whole repertoire of ways of moving from one place to another: walking, running, jumping, scrambling, swimming. Horses have several gaits: walking, trotting, galloping. Birds can walk or fly. Here, vicarious functioning is very sophisticated because it is not only about replacement but also about adaptation of a mode of locomotion to environmental conditions or to the urgency of reaching a goal. This is *adaptive* vicarious functioning.

Elsewhere I have described some of the amazing mechanisms that underpin walking. For example, the creation of mobile reference points to control coordination of the limbs with the stabilized head; the similarity of the laws that govern locomotor trajectories with those that control the hands and gestures; the mechanism by which gaze anticipates a trajectory; the stereotypical character of trajectories; and so forth.[45] Walking is a highly hierarchical function, but also profoundly vicarious because it is also astonishingly flexible.

The base rhythm is produced by networks of neuronal oscillators situated in the spinal cord. These networks can reconfigure themselves to induce different locomotor patterns (walking, running, jumping, and so forth). In addition, the centers for triggering and beginning these different movements are located in the mesencephalic reticular formation of the cerebellum and the basal ganglia (which is why patients with Parkinson's have trouble taking a first step: the basal ganglia lack dopamine). These centers integrate multiple data regarding the environmental context, the intention, and the goal of the action under way. Moreover, walking is also controlled by the emotional state of the walker, their social status, and their culture. The undulating walk of a seducer played by Marcello Mastroianni is very different from the step of a soldier during a military parade. And joyous walking is different from sad walking.

Vicariance in walking may be dependent on sensory cues. Yet the human brain is capable of directing the locomotor path even if sensory input is absent. You can get around your apartment even in the dark. One of the most adventurous outings I remember from my scouting days was when we had to find our way in complete darkness. Instead of visual cues, humans rely on a mental image of the surrounding space and update it. To study this phenomenon, researchers have devised tasks called "locomotor pointing."[46] Subjects must walk, eyes closed, while pointing their finger toward a fixed target. It is easy to do this experiment yourself. It is a fine example of "spatial updating." Using mental imagery as a substitute for the real world is a powerful mechanism of vicariance.

In this context, vicariance refers to the creation of a genuine double by the imagination. Recent data show the similarity of mechanisms of path control both for actual walking and when the walking is imagined. For example, there is isochronism between actual walking and imagined walking.[47] Moreover, when you turn while walking around an obstacle, the brain areas activated are similar whether you actually execute the turn or you imagine turning.[48] Finally, imagined walking causes gaze movements similar to those that occur during actual walking. The brain creates a vicarious virtual walker.

Mental Imagery: A Mechanism for Constructing Future Scenarios

The neural basis of mental imagery is worth a closer look. Some regions of the brain are particularly called into play. One of them, the precuneus (in the parietal lobe), is involved in a great variety of tasks that use mental imagery: episodic memory, changing point of view, agency, awareness, and self-reflection. That is why it is called the "mind's eye." The precuneus is divided into two subregions: the anterior region, which is especially involved in mental imagery, and the posterior region, linked to aspects of episodic memory that are associated with mental imagery. If we examine the areas of the brain that are involved in perceiving or in imagining a tree, we find similarities and differences. One part of the network is shared, but coverage is more pronounced in the frontal and parietal areas than in the occipital and temporal areas.[49] Deficits in imagery can be independent of deficits related to visual perception.[50] Two different networks are active. One is involved in processing "abstract" information; it activates mostly left anterior and superior language areas. The other network involves processing "concrete visual" information, and activates right posterior temporal regions that are involved in working memory. "Working memory" refers to the capacity of the brain to briefly memorize a bit of information, for example, a telephone number. This capacity allows the brain to combine information independent of its source.[51]

The neurologist Wilder Penfield identified the brain areas involved in remembering past experiences.[52] "Experiential hallucinations" is the term he gave to the memories he evoked by electrical stimulation of these areas in people with epilepsy. In fact, they were most likely not hallucinations but genuine memories of past experiences.

Mental Time Travel

Recent data from magnetic resonance imaging suggest that mental simulation of future events activates more or less the same networks as memory of past events.[53] In fact, a network of brain areas involved in the creation of future scenarios has been identified. It comprises, among others, the medial prefrontal cortex, the medial temporal lobe, the retrosplenial cortex, the posterior cingulate cortex, and the inferior parietal lobule in connection with the hippocampus. It has also been proposed that the networks involved in future scenarios are specialized in the same way as the networks for different memories. They are said to be "in register," which facilitates the connection between past and future.[54] These networks form the basis of the "prospective brain."[55] However, this theory remains debatable because other studies show that the networks that process imagination of future events based on past memories are different from those that process episodic memory of past events.[56]

A major tool of vicarious creativity is the link between memory of the past and scenarios of the future. The human capacity to remember the past and to predict the future by "mentally traveling in time," a genuine round-trip between two opposite poles, involves the regions of mental imagery noted above. Here, Edmund Husserl's notion of "retention-protention" is relevant: "The protentional horizon of intimate awareness of time oriented toward an imminent event is structured by a turning inside out (*Umstülpung*) of the retentional horizon of the immediate past." In this way, our expectations (though inevitably inadequate) are still never completely indeterminate. As the retentional horizon is eroded by determinations, the protentional horizon fills in the gaps of our expectations, and this double movement preserves the stylistic coherence of time consciousness. "The style of the past," writes Husserl, "is projected onto the future." This endows the coming event with a certain familiarity that enables us to "welcome it with open arms."[57]

There are two distinct kinds of mental time travel: "absolute" mental time travel, which corresponds to self-localization at different instants, or "positions," in time (past, present, and future), and "relative" mental time travel, which situates the self in relation to past, present, and future events. The capacity for mental time travel is specific to humans and may be one of the reasons for our evolutionary success. Finding alternative solutions, which is a characteristic of vicariance, rests on this back and forth between past experience and anticipation of the future. What is known as episodic memory is thus one of the tools of vicariance. It does not suffice to conjure up recollections from the great dustbin of memory. One must evaluate the relevance of past events for predicting the future and choosing a solution.

But even though episodic memory enables us to recall past events, these memories may not be put together correctly. The result is confusion: judgment becomes clouded and prediction unreliable. Harvard psychologist Daniel Schacter has studied memory errors in the case of mistaken evidence leading to confusion following the 1995 attack on the federal building in Oklahoma City.[58] The limits of memory have also been documented by Armin Schnider, at the University of Geneva, with respect to "confabulators."[59] These subjects often exhibit damage to the orbitofrontal cortex and can neither update nor select past events that pertain to the present and are thus important for constructing future scenarios. For example, one hospitalized female patient insisted that she had to go bottle feed her infant son. The woman did have a child, but he was thirty years old.

I would like to venture the following paradoxical hypothesis, for which I do not have proof: *A memory deficit may be a powerful tool for creating possible worlds and devising solutions.* The limits of memory may be used by the brain in the process of creative imagination. For example, a fragmented memory could provide the brain with material for constructing not only dreams but also novel scenarios. Vicarious functioning might then use this material to invent new pathways. After all, one of the amazing properties of living organisms is their ability to turn their weaknesses into advantages. This touches on one of the great mysteries of evolution. The implacable battle for survival based on competition simultaneously gives rise to new properties that life exploits to survive, such as the fact that the weak arouse a help response in humans and other animals, or the mystery of compassion or ethics.

Memory of the past is fundamental in creating future scenarios and in selecting vicarious solutions. But not all memories are equally important for influencing decisions and choices. Research on autobiographical memory, for example, has revealed a key period in the construction and maintenance of identity. This slice of life, called a "reminiscence peak," corresponds to memories encoded in adolescence and young adulthood, that is, between roughly eighteen and thirty years of age. This period involves personal semantics and episodic memory. Brain imaging has identified the brain regions that come into play.[60]

Is the Brain an Optimist or a Pessimist?

Emotion enters into this temporal dynamic between past and future. For example, the capacity to imagine vicarious scenarios is influenced by a person's tendency toward pessimism or optimism. The area of the brain that matters most in attributing a positive or negative value to a stimulus, or in deciding whether it is dangerous or benign, is the amygdala. Researchers have shown that, all other things being equal, its activity is greatest in pessimists. No doubt this corresponds to its decisive function in inducing fear. Indeed, the amygdala is the area of the brain that detects danger and very rapidly triggers one of three fundamental fear responses: fight, flight, or paralysis.

The anterior cingulate cortex is closely associated with optimism. It has many functions, but one particular region, a "motor" region, located below the supplementary motor area, is most likely involved in the control of movement. Another, "cognitive," region, farther forward, is involved in cognitive tasks. It is activated by actions related to conflict, for example. Finally, the anteriormost, "affective," region is involved in regulating emotion. Detailing the many recent discoveries made on this subject is beyond the scope of this book. What is important, however, is that emotion is a fundamental component of vicariance. Étienne Koechlin recently proposed that the anterior brain is involved in three major aspects of emotion in decision making, and consequently in the diversity of behaviors for dealing with a particular situation: the frontal cortex is involved in a person's motivation (I want to do this or that); the orbito-frontal cortex in the attribution of value (I prefer this); and the prefrontal cortex in decision making (I decide to do this).

Vicariance and the Variety of Conscious Processes

The concept of vicariance is undoubtedly also compatible with multiple consciousnesses. The neurophysiologist Semir Zeki writes: "Attempts to decode what has become known as the (singular) neural correlate . . . of consciousness . . . suppose that conscious-ness is a single unified entity, a belief that finds expression in the term 'unity of consciousness.' . . . I propose that the quest for the [neural correlate] will remain elusive until we acknowledge that conscious-ness is not a unity, and that there are instead many consciousnesses that are distributed in time and space."[61] The neurophysiologist Al-fred Fessard identified four brain networks involved in four kinds of conscious processes:

1. The *corticocentric* process, exclusively involving links between the regions of the cerebral cortex: this is the recent theory of Stanislas Dehaene and Jean-Pierre Changeux.
2. The *cortico-subcortical* process, involving the networks linking the cortex to the basal ganglia: this is the theory supported by Rodolfo Llinás.
3. The *centrencephalic* process, combining the two preceding processes: this theory was proposed by Wilder Penfield.
4. *Primary consciousness*, an elementary form of subcortical consciousness found, for example, in anencephalic infants. This may also correspond to what neuroscientist Antonio Damasio calls the "proto-self."

Consequently, one shouldn't speak of consciousness as a unified entity in opposition to the unconscious.[62] Using the singular to describe the manifold capacities of the human brain would be to deny the extraordinary adventure of progressive development of these properties during evolution. Each of these faculties—awareness, memory, attention, anticipation, vision, perception, the unconscious, imitation, and so forth—in reality harbors multiple levels of complexity. They continually interact and are called into play depending on the task, the context, and the individual. Multiple consciousnesses are essential to the theory of vicariance that I wish to propose in that they permit the creation of multiple solutions for resolving a problem. This is not the place to touch on the fascinating case of hypnosis. But my group has done some limited studies on the problem of different susceptibilities to hypnosis; this manipulation of states of consciousness induces very different reactions in different subjects.[63]

3

The Personal Body, Self, and Identity

In addition to the relatively trivial issue of replacing the priest with a vicar during a religious service, the notion of vicariance raises the more profound question of duplicating the person of the priest, including his skill, identity, and function. In fact, the major issue at hand is that of a person's identity and relationship with others. Today there is a veritable explosion of thinking, scientific papers, and various experiments on the question of identity and its disorders, on multiple opinions, shared emotions, and empathy. New disciplines, such as social neuroscience, attempt to determine the biological basis of social psychology, respect for moral standards, the origin of violence, and so forth. The discovery of mirror neurons, sometimes applied in other contexts, has narrowed the distance between self and others and raised pointed questions about our independence vis-à-vis others.[1] The problem of identity is also posed by new techniques of brain stimulation in what is called the brain–machine interface. This consists of implanting electrodes in the brain of a patient, for example, someone who is paralyzed, that enable the patient to control a robotic aide. However, the first attempts at this technology showed that patients feel uncomfortable, as if someone else were directing things in their place (this feeling of loss of control is also typical of people with schizophrenia).

Today the theme of the double, which I discussed in *Emotion and Reason,* and in Chapter 2 of this book, has taken on a new dimension with the discovery of the neural basis of split personality and of out-of-body experiences, which had previously been unexplained. In the sections that follow I will refer to various notions about the body. These include the physical body, the anatomical body, the personal body, the body schema, and body image. I do not purport to explain all these notions, which are discussed elsewhere. My aim is to shed some light on the concept of vicariance by showing how these various categories reflect the basic flexibility and the possible duplication of representations of the body.

The relationship between the body and the world invites reflection on the notion of self and identity. This subject is huge, and I will only touch on several useful aspects for our contemplation of vicariance. There is not a unitary, fixed self, independent of the vicissitudes and uncertainties of life. The self poses the same fundamental question that vicariance does: the relationship between the permanence of the self (a common conceit) and its changeability, its flexibility. This subject has of course been thoroughly examined by philosophers and psychiatrists (varieties of humors, a different interpretation of emotion, no doubt attest to this changing self), but here I would like to bring it to the fore.

The difficulty with an empirical approach is evident in the writings of philosophers and psychologists. Maurice Merleau-Ponty refers to this subject in defining what he calls the "essence" of the individual: "One is led to the idea that there exists a general structure of behavior for each individual which is expressed by certain constants of conduct, of sensible and motor thresholds, of affectivity, of temperature, of blood pressure . . . in such a way that it is impossible to find causes and effects in this ensemble, each particular phenomenon expressing equally well what one could call the 'essence' of the individual."[2] This idea, in addition to the body schema made up of the neuronal networks simulating the properties of the personal body—what neurologists call a "body image"—is also discussed by the psychologist Albert Michotte, who wrote extensively about what he called the "phenomenal self."[3]

Vicarious Identity: Multiple Identities

The problem of identity has given rise to myriad theories and controversies.[4] Recall Aristotle's question: *Ti esti?* (What is it?) The Aristotelian response is: X is a man, an apple tree, and so forth. This individual response, writes philosopher David Wiggins, "implies also a *phusis* ('nature'). The *phusis* is the mode of being and acting of a group of things that share among themselves a distinctive source of movement and change, and thus their own principle of activity."[5] Yet, according to ethno-psychiatrist Tobie Nathan, "In African cultures, a person's identity is hidden, . . . and must be defined as the action that this person is destined to exert on the world."[6] The eighteenth-century French philosopher Pierre Maine de Biran also dissociated the notion of existence from that of substance in relating it to action. For Maine de Biran, to say "I am" is to say "I want, I move, I do." Here, agency is central to the definition of identity.

Most surprising is the paradox between invariant identity and the multiple identities possible for the same person. You might say that the paradox is that identity is both invariant and vicarious. John Locke, who was taken with the idea of invariance, wrote: "[A person is] a thinking intelligent being, that has reason and reflection, and can consider itself as itself, the same thinking thing at different times and places."[7] The philosopher Paul Ricoeur, like many others, attempted to determine the relationship between vicariance and invariance. He wrote: "Must one not, in order to make oneself open, available, belong to oneself in a certain sense?" and "For my flesh appears as a body among bodies only to the extent that I am myself an other among all others."[8] As for novelist Gustave Flaubert, he identified with his heroes: "Madame Bovary is me," he is said to have written.

Nonetheless, despite the apparently invariant character of identity, for Ricoeur, the "same" appears in several guises: (1) identity as sameness or *idem*-identity (*mêmeté*) (*idem* implies permanence in time)—the German equivalent is *Gleichheit;* (2) identity as *ipse*-identity (*ipséité*) ("*Oneself as Another* suggests from the outset that the selfhood [*ipséité*] of oneself implies otherness to such an intimate degree that one cannot be thought of without the other"[9]). The Latin *ipse* is *self* in English, *selbst* in German, *se* in Italian, and *sí mismo* in Spanish; (3) *narrative* identity: "It is indeed in the story recounted, with its qualities of unity, internal structure, and completeness which are conferred by emplotment, that the character preserves throughout the story an identity correlative to that of the story itself."[10]

The diversity of identity content is also emphasized by Shaun Gallagher.[11] He distinguishes the "minimal self," a self "devoid of temporal extension," from the "narrative self," which involves "personal identity and continuity across time," based on two modalities: "what others say about me" and "stories I tell about myself." In particular, Gallagher distinguishes a unique self, the synthesis of all cognitive and emotional components, from a self that is fragmented into a myriad of different modalities.

These research efforts demonstrate several characteristics of identity. Over a lifetime, body identity changes. Moreover, there are many levels of identity. Identity is based not only on language but on past action (episodic and retrospective memory) and, especially, future action: this is the concept of "power of action" (from Duns Scotus, the medieval philosopher-theologian). Identity is thus a dynamic process that connects memory of the past, the creation of the present, and prediction of the future, which gives rise to the idea of mental time travel.[12] Perhaps some people in search of an identity, who cannot find it in the present or the past, search in the future for an identity that they will never find—a Holy Grail. Such is the long trip toward the kingdom of Brother John in Umberto Eco's *Baudolino*.[13] (I will discuss utopia as an ultimate form of vicariance in the Epilogue of this book.) Identity is also founded on intersubjectivity; it is profoundly linked to our relations with others.[14]

The process of constructing identity is thus, in my opinion, a vicarious one. The round-trip that it involves between present, past, and future enables it to take multiple, vicarious forms that preserve an invariance whose biological basis we are still far from understanding. I suspect, as psychologist Philippe Rochat proposes with respect to infants, that this invariance derives fundamentally from the intersubjective, narrative, and social nature of the construction of self. But I might add, in line with my idea of a brain essentially turned toward the future, that this invariance is mainly due to the fact that we ourselves are turned toward the future. Gilbert Simondon writes: "Being does not possess a unity of identity, which is that of the stable state in which no transformation is possible; being possesses a transductive unity, which is to say that it can diphase itself in relation to itself; it can overflow out of itself from one part to another, beginning from its *center*."[15]

Simondon's observations illustrate several essential elements of my thesis: vicariance is animated by projection into the future, and it requires not only delegation and duplication (the vicar), but a "decentering" that enables a change in perspective, or point of view, that spurs creativity. Thus, this book includes chapters on manipulating spatial points of reference and the difference between sympathy and empathy.

In short, it is the need to anticipate more or less equivalent (vicarious) scenarios that constructs an identity. This identity is the result of the tension between the diversity required for vicarious scenarios and the indispensable unity of the agent that can implement them.

This illustrates the opposition between the universal and the particular that I mentioned in the Introduction.

The concept of vicariance thus also concerns changes or substitutions of identity. The anthropologist Philippe Descola cites an interesting example of replacement of one identity by another. He describes a tradition among the Achuar, a tribal subgroup of the Jivaro people of the Amazon. Each group has a limited number of identities. When a member of the group dies, one of these identities is lost and replacement is effected by cutting off the head of one of their Jivaro neighbors. The head thus retrieved is mummified: "The aim is not to create a trophy to prove victory over an enemy but to capture the singular identity characterized by a face, itself singular."[16] To my mind, it represents an original form of socially vicarious identity.

Evolution provides many examples of vicarious changes of identity for reasons of security, predation, or reproduction. These include the seasonal alterations of anatomy in birds and mimicry in animals.

The King's Two Bodies

An example of double identity that is not the result of pathology but of a social choice intended to separate social functions is that of the king's two bodies. The crown lawyers for Edward VI of England spoke of the king's body in this way: "[T]he King has in him two Bodies, viz., a Body natural, and a Body politic. His Body natural . . . is a Body mortal, subject to all Infirmities that come by Nature or Accident. . . . But his Body politic is a Body that cannot be seen or handled, consisting of Policy and Government. . . . [T]his Body is utterly void of Infancy, and old Age, and other natural Defects and Imbecilities."[17]

Multiple Identities: Strength or Weakness?

Take as a first example the question of identity, which is at the heart of many social problems. We are currently witnessing a proliferation of identities. In the past, a person had a limited number of them, such as citizen, head of family, member of a political party; one had an occupation and a skill. Today, identities are proliferating because of social openness, global migration, and the growing number of institutions or groups one might belong to.

Identity is also associated with breathtaking revolutions in the nature and organization of work and its "management." Indeed, this was the focus of a symposium on work, identity, and occupations that I organized with Yves Clot at the Collège de France; it became obvious that the major transformations that have occurred in the content of work and in hierarchical relationships have had a major impact on the identity of agents and workers.[18] Yves Clot describes the disruption caused within the French postal service and news agents by the introduction of excessive commercialization: agents were told that if a customer presented a badly wrapped package, instead of helping them to close the package, the agent should propose a postal service product that the customer would have to pay for. Agents saw themselves transformed from public servants whose mission was to help the public into salespeople whose mission was to bring in as much money as possible. Financial productivity had replaced the mission of prosocial behavior.

I have myself observed that management models and theories are very far removed from the reality of work and dramatically change the identity of the worker. For example, the notion of "process" ties an employee to a set of rules and procedures that suppress all individual initiative and are dictated by financial criteria to the detriment of the quality and the creativity of work. The time constraints of the rhythms of production (so many pieces per hour), the limits of intervention (so many minutes to speak to a client by phone), the lack of autonomy, and changing skill profiles conjure up in the West the depressing memory of Taylorism and, in Russia, of Stakhanovism. Today, millions of workers throughout the world suffer from musculoskeletal disorders and associated cognitive and temporal stress because of work requiring standardized and repetitive movements.

Whose Name?

Another way of changing identity is to change one's name. But a name is not just a label. Tobie Nathan writes that "in certain African societies, . . . the naming process implies an entire group, its history but also perhaps above all an individual's path as it is being accomplished, what I will call his project. . . . To say of someone that his name is so and so, certainly isn't to take a stand on that person's specific essence, but it is always a prediction on the state of the world owing to that person's existence."[19]

Changing one's name is to change one's destiny: "A person is not a natural object, but a cultural artefact."[20] This is what happened after the Holocaust to the "hidden children" whose parents chose to change their children's names in the hope that their children would not be identified as Jewish and would no longer be persecuted or discriminated against. It is a curious feeling to live under a false name, as it were, and to be one's own vicar and that of someone else at the same time. One must come up with an origin connected to the name, but one that is not true: consequently, one ends up having to lie in perpetuity. What a fate to have to lie forever about one's identity! A fraudulent vicar for someone who, moreover, one does not know and who has perhaps not agreed to have a double! How odd to take the place of oneself. One hidden child described the situation thus: "I was not capable of being beautiful. Because to be beautiful, I would have to be *capable of putting myself in the place of another*. And I simply could not do that. In other words, to be able to see that I was beautiful, I would have to see myself; yet I did not see myself. I did not think I was ugly. I had no image of myself at all. I was a face without features, an external form that had no reality for me."[21]

This is constrained vicariance, which destroys identity and profoundly alters social relationships. You might ask whether digital avatars, these virtual personalities in computer games that substitute for players and represent them in the virtual world, may also paradoxically perhaps restrict a player's socialization in the real world through a mechanism of depersonalization analogous to the one at work in the hidden children. I will come back to this point later on.

The Virtual Body: The Fregoli and Capgras Delusions

Evolution has endowed us with various ways of multiplying our identity and our body, based on the property of emulation that I mentioned above. The actual human body and brain are replaced by virtual versions. Merleau-Ponty employs the notion of "virtual body," which to a certain extent echoes the concept of the double that I discussed in *Emotion and Reason*: "This virtual body ousts the real one to such an extent that the subject no longer has the feeling of being in the world where he actually is, and that instead of his real legs and arms, he feels that he has the legs and arms he would need to walk and act in the reflected room: he inhabits the spectacle."[22] It is interesting to see the term "reflected room" used in the language of philosophy.

In psychiatric pathology, in addition to out-of-body illusions (autoscopy and heautoscopy), several types of disorders of identification or creation of virtual identities have been described that are relevant to vicariance.[23] The most famous is Capgras syndrome. An example is a child who falls off his bike and returns home. Seeing his father, he says: "You look like my father, but you are not my father." People afflicted with this disorder believe in the existence of doubles of people with the same physical appearance but not the same identity. It is a syndrome of hypo-identification. Some patients believe that copies of themselves exist, but with minds different from theirs.

Another syndrome, called Fregoli syndrome, is the opposite of Capgras syndrome because it involves hyper-identification.[24] A person with this syndrome believes that the identity of the familiar person is unchanged, whereas their physical identity is totally different. Patients with Fregoli syndrome may interpret the mind of the person they erroneously identify as that of anyone except the person he or she is. Finally, intermetamorphosis syndrome, described by psychiatrists Paul Courbon and Jean Marie Tusques in 1983, produces in patients the illusion that others have undergone a profound change in identity, both physical and psychological.

The neural basis of these disorders is still not well known because the forms are so varied. It is assumed, for example, that an inter-hemispheric disconnection between cortical areas allows each area to establish representations independent of a person, place, or event. Consequently, if one cerebral hemisphere cannot explain the perception received by the other, the patient processes this difference by confabulating around the lived experience. This has been shown in patients presenting with callosal disconnection ("split brain").

The Avatar Revolution

We could say that we have an internal vicar—the body schema that is implemented in the structures of the neural networks and signal processing properties of the brain. This is a biological double homologous to the human body. Thanks to internal models, it possesses the properties of the physical world around us. The pathologies I described earlier concern disordered interpretations of the real self. Now a spectacular revolution in information technology has been taking place for several years. Computer scientists have created avatars, that is, virtual personages. These are just computer programs, but they can be shaped and given very diverse properties, characteristics, personalities, and intentions.

Current cinema makes generous use of these avatars, which are veritable hybrids because their movements are often produced by using digital imaging techniques to capture real movements of humans who also lend the avatars their voices. In Second Life software, and many computer games, you can construct one or several avatars of yourself and live another life. In certain games, the avatar of the player is the central character. In this way, the game is played by delegation, which is virtual vicarriance. The artist Claire Sistach immersed herself in a video game with several avatars. One of the avatars regularly lost. The player's identification with the unfortunate avatar caused her to become depressed. So she created a second avatar who won, which made Sistach feel very happy and optimistic. It is remarkable that it took only a mouse click for this artist to feel sad or joyful *herself*. Virtual avatars are thus genuine "embodiments" that are internalized to the point that their creator takes on their emotional life. Vicariance, in this case, consists of real identification with the virtual creature.

A recent experiment also shows the evocative strength of substituting an avatar for a real person. The group of François Garnier, a professor at the École Nationale Supérieure des Arts Décoratifs in Paris, created a collective virtual visit to several museums in Berlin. Here is the brief description of what they did. Imagine that you are sitting at home in front of a computer screen displaying a museum gallery that contains several works of art. In the gallery you see an avatar of yourself, that is, a figure that talks and thinks as you do. In the same gallery, you also see one or several other avatars. These are the avatars of people who, like you, are at home, sitting in front of their computers, possibly on the other side of the world. You can ask your avatar to go stand in front of a painting and discuss it with the avatars of the other people. You are creating a virtual visit to the museum by interacting with people that you do not see except through their avatars. The most remarkable thing is that when you are in this situation, you really have the impression of talking with the other people represented by their avatars. Teleported into the virtual world, you have the feeling that the other real people are there, too.

The capacity of the human brain to imagine itself, together with its feeling body, in a virtual world, is a major discovery that goes far beyond the well-known feeling of entering a fictional world when reading a novel. The brain is not only a creator of worlds; it can also enter into these worlds, as in a dream. Étienne Pereny, a digital imaging specialist, has written: "Therefore these worlds, these objects, their interconnection, and this generalized access through the network can be considered a new type of technological milieu, a milieu of human hybridization[25] that alters humans' relationship with both machines and the world. What appears to characterize these new technological assemblies is the intimate, broad relationship that they establish between the real and the virtual, more specifically between the material reality plane and the immaterial reality plane, but nonetheless habitable and practical."[26] Pereny and his colleague Étienne-Armand Amato, also a specialist in digital worlds, have proposed a distinction between the avatar of the classical virtual world that performs tasks, but where there is no identification with the player (a vicar in the strict sense of the term), and that of a video game that "instanciates" the self of the player.[27] In speaking of "instanciation," Amato refers to the notion of "instance," which, in computer science, designates the variant obtained from a programmed model that is updated on the screen, for example, a personage or a vehicle. Depending on the case, by identifying with this cybernetic creature that is endowed with various behaviors, the player may perceive the avatar as his double to the point of fusing with it, or, to the contrary, may perceive it as an extension of himself that can be used as a tool.

Pereny writes of this new vicar: "The intentionality of the human being is transformed by externalization, by delegation of his desires and actions to the program or the device. The instanciation system in video games causes the player to enter into symbiosis, into constitutive coexistence with an impression of reality, both with the information and with the intentionality written into the program."[28] He also emphasizes the role of movement as a trigger and supporter of this hybridization. "For these new images are, in part, mental images generated by the brain and produced by gesture. The direct cognitive fruits of motor involvement of the subject, they proceed precisely as Berthoz describes in *The Brain's Sense of Movement* by unleashing sensory perception through the simple movement of a command."[29]

These specialists in media and interactive images have in fact proposed a distinction with respect to avatars, between a "hyper" logic based on commands given by clicks and hyperlinks, typical of the Web, and a "cyber" logic based on control and coupling, as in video games. Only the latter enables sensory participation, with various degrees of investment, in virtual worlds.

However, the risk posed by video games and the use of avatars is not negligible. I was told a horrible story of two young boys who were playing a video game together in which they were supposed to kill the first person they met. They had so completely experienced the scenario as real that one evening they went outside and killed the first passerby they encountered! It is obviously worth asking how much of a role is played by the video game itself, and how much by the personality of the players. This is a question for psychiatrists to answer, if they can.

Humanoids and Robot Companions

For many years, humanoid robots were mainly designed in Japan.[30] Recently, in France and elsewhere in Europe, several groups have entered the field. At the Collège de France, an annual chair was designated for robotics. Jean-Paul Laumond was one of the holders, and together we organized a symposium on neurosciences and robotics to mark the convergence of our disciplines. In this area, I will discuss two subjects that touch on the notion of vicariance: *humanoid robot companions* and *redundancy*.[31]

A new form of vicariance is currently emerging in the form of a humanoid robot endowed with a very limited level of intelligence: the robot companion.[32] As the name suggests, this entity is more a companion than a vicar for a human subject. Quite a while ago now, Roland Siegwart, a roboticist in Zurich, displayed at an exhibition eleven robots equipped with wheels and eyes that accompanied people on their visit. The challenge is to give these robots properties compatible with interaction: modes of language, perception, action, memory, and decision. They can take human forms, but also functional forms for applications related to work, education, and rehabilitation after brain damage. They must be highly adaptable, and able to take account of differences between cultures. For example, we are working on projects with Japanese colleagues at Waseda University.[33] In Tokyo, they have built a robot named Wabian that expresses emotions and is beginning to interact with humans. In Paris, Bruno Maisonnier, of Aldebaran Robotics, has created the Nao robot and its big brother Roméo. Intended to serve as play companions or to help their owners in performing daily tasks, such robots are proliferating and will one day invade our lives.

Today, efforts are ongoing to fabricate humanoids controlled by the brain through encephalographic recordings of cerebral activity. They will be artificial "vicars" directly connected to the brain of the "priest/subject" they will replace. These companions will be our slaves. They will perceive, feel, act, and perhaps decide in place of their "master." The master will only have to think of something ("I would like a glass of orange soda") for his brain activity to be decoded by the robot, who will head for the refrigerator. This scenario is criticized by detractors who say that a speech decoder would be sufficient. But beware of false prophecy: a robot nurse was on the cover of *Science et vie* in 1910 . . . and we are still waiting.

Applications for these robots will not be limited to assisting patients. Some humanoids and intelligent robots will be specialized in managing natural disasters (nuclear accidents, fires, search-and-rescue missions following earthquakes).

Humanoid robots are not the first creatures that humans have used to replace themselves. In fact, animals were the first vicars. Dogs replaced shepherds to corral herds, and in so doing developed remarkable skills and knowledge that surpassed that of their masters. These dogs are able to understand and manage the feelings of the herd. It is as though the vicar knew the parishioners better than the priest! In the Pyrenees, a shepherd who practices transhumance told me that a dog doesn't have to control the herd on every side. It's enough to work the sheep on one side; through mutual attraction, the herd spontaneously organizes on the other side as if it were a single mass. If one dared, one could say that slaves were living humanoid robots, and that the mercenaries employed even today in war are men used as robots for killing. Today, the contempt of rich countries has not changed very much; it's just that now it consists of regarding the populations of poor countries as mere consumers.

One challenge to introducing humanoids into civil society is the diversity of cultures. For example, in Japan, when you meet someone, you bow slightly to greet that person, whereas in the West you shake their hand. Consequently, the Wabian robot built at Waseda University by Atsuo Takanishi is programmed for one or the other of these behaviors, based on its interlocutor. It's a matter here, of course, of one of the most elementary elements of an interaction. Recognizing an Asian as opposed to a Westerner might involve the shape of the eyes or vocal characteristics. But real vicariance requires going much further. The humanoid robot would have to understand the intentions of the human, be able to judge the success or failure of the task it has been assigned, and inspire confidence. In

Japan, artificial creatures are perceived as benign and are endowed with almost human faculties. In Europe, it is more difficult to get people to accept a humanoid as a robot companion. This may be due to the story of the golem, a creature of clay that was developed to defend the Jewish community in Prague but then turned on its people, or of Dr. Frankenstein, who was tormented by his creation. Other explanations are obviously possible, such as the fact that a dominant idea in the West is that humanity was created by God and cannot itself be endowed with the power of creation. I will leave it to my eminent colleagues to speculate further on this subject.

Another difficulty stems from the effect known in robotics as "Uncanny Valley."[34] The problem is the following: in general, the more human a humanoid or artificial creature appears (say, an avatar in a film), the more it is accepted and becomes familiar. However, beyond a certain threshold of resemblance, an inexplicable phenomenon occurs. The character arouses mistrust and rejection. If the character continues to be perfected, it may eventually become acceptable again. As a result of this "valley" in perception, roboticists are currently giving humanoids features that are nearly, but not completely, human, so as to remain within the limits of acceptability. One possible explanation, supported by recent studies, is that the brain is troubled by the incongruity between the human features of humanoid forms and their unnatural movements. In other words, the brain cannot reconcile the facial features with the motor behavior. I have long been recommending to roboticists that they factor in the "laws of natural movement" (which I summarized in *Simplexity*). It would be interesting to know whether, as recent studies suggest, taking these laws into account causes the valley to disappear or diminish.

One of the many properties given to humanoid robots is that of *redundancy*. I have already mentioned this term among the principles of simplexity applied to living things. In a particular context, a redundant system can compensate for the shortcomings of a process by replacing it with another. In the fields of artificial intelligence and operational research, the notion of "substitution or compensation," which underpins the concept of vicariance, implies finiteness, that is, a limited number of choices, and at the same time a discrete structure of the "finite state machine" type. Robotics, too, makes use of "decisional architectures."[35] Transposing these notions to the finite domain, which is that of robotic control, leads naturally to the idea of functional redundancy, that is, that the habitual movement that a robot must execute to pick up an object can be modified at leisure based on context: making a detour if the robot encounters an obstacle, say, or compensating for temporary motor defects, and so forth. This functional flexibility highlights the relationship between redundancy and vicariance. To take advantage of redundancy, what matters is that the motor capacity of the machine be "superdimensional": the dimension of motor space (that is, the number of degrees of freedom) is greater than the dimension of the task to be done (for example, placing the hand on an object). It is all a question of geometry: motor space must be related to action space, the former having a much larger dimension than the latter. Developing these geometries is part of what roboticists do. They must master the motor redundancy of humanoid robots. For example, the team of Jean-Paul Laumond has shown that the humanoid robot HRP2 can "call on" its legs to grab an object that is beyond the reach of its hand, without having had the legs specifically programmed to do so—a capacity known as embodied action.[36]

The emphasis on standards in technology can lead to the destruction of vicariance. Whereas there used to be a thousand ways to repair a car, for newer cars equipped with electronics, there is now only one: you replace the entire defective system, just as you would discard a mobile phone, because repairing it is more expensive than buying a new one. Roboticists working on humanoids would do well to keep this in mind. One challenge for them is how to endow the robot with the capacity for recombining systems to replace defective mechanisms, as the nervous system does.

4

Vicariance and Changing Perspective

> In the modern sense of the word, "existence" is
> the movement through which man ... involves
> himself in a physical and social situation which
> then becomes his point of view on the world.
> —MAURICE MERLEAU-PONTY

Does Vicariance Depend on Multiple Spatial Reference Frames?

I have shown that the choice among possible scenarios, an essential property of the brain for inventing vicarious solutions, requires adopting a new perspective and changes in points of view. To discover other ways of accomplishing a given task, one must be able to envisage it while changing perspective. Building an identity in a social context also relies on changing perspective since, in the words of Paul Ricoeur, it entails being "oneself as another" and since, according to Gilbert Simondon, individuation requires that one "decenter" oneself. Interaction with virtual avatars also means changing perspective, and many virtual games require the capacity to change places with a mental agility that young people today seem to possess in spades.

Changing perspective or points of view requires that the brain change "spatial reference frames." Humans are continually changing their reference frame. I call this "manipulating spatial reference frames" even though the operation is a mental, not a manual one.[1] I believe that this rich repertoire of potential reference frames enables the vicarious functioning defined earlier. I will provide several examples of what I mean by changing reference frames.

In the sensory domain, vision supplies a frame of reference for the vertical and spatial orientation of the environment. The vestibular system provides a point of reference linked to the head and uses gravity to evaluate the orientation of the head in space. Tactile and proprioceptive systems give information about the space around the skin. The soles of the feet, which establish a point of reference with respect to the ground, are thus a veritable "tactile retina." Depending on the context and the task, the brain can decide to take as a point of reference the feet on the ground, the head (if the ground is unsteady), or even the simple touch of a finger on a stable part of the environment. A touch of the finger on a fixed point, a wall, say, can induce the brain to take that point as a reference.

When we set the table for dinner, we place the silverware and plates within the space of the table. We are situated within the reference system of the table, and it does not matter whether the table is localized in the dining room or in the kitchen. In placing the silverware, we are working within the reference system of the plates. And when we sit at dinner eating, the plates, the silverware, and the wine bottle we grasp become located in our egocentric, body reference frame.

Consider now the example of dancers. The choreographer Noa Eshkol proposed describing the same movement with respect to three different frames of reference: one linked to the body (bodywise, or egocentric), another to the partner (partnerwise, or heterocentric), and a third to the environment (environmentwise, or allocentric).[2] For example, these three systems of reference correspond, respectively, to the dancer turning her head to the right of her body, toward her partner, or toward the back of the stage.

Aside from various reference systems for relationships between the self and others, the brain is able to invent worlds in which it can imagine being. When you turn on your cellphone or your tablet and begin to navigate through a social network, you are suddenly teleported into imaginary spaces and points of view. You can also imagine the position of this book vis-à-vis your eyes, the door to your home, the nearest post office, or the city of London—examples chosen to avoid your having to situate the book with respect to the star Aldebaran or the constellation Orion.

At school, teachers have opinions about students, but they can also ask what the students think of them; at the theater, an actor might wonder what the audience is feeling. In social life, people frequently change their reference frame by taking others as their reference. As I will explain later, this is the difference between sympathy and empathy.

Sometimes it is necessary to use these different points of view simultaneously. Suppose that you want to cross a street that has no crosswalk. If a car comes, you have to control your steps while deciding whether or not to cross, all the while putting yourself in the place of the driver to guess her intentions. You must adopt, successively or simultaneously, several points of view and situate yourself within several reference frames: that of your body, that of the car with respect to the street, and that of the driver. When a striker in soccer takes a penalty shot, he positions himself within a reference frame that relates his foot to the goalie cage, but at the same time he has to build a mental image of the way the goalkeeper sees him. This requires a simultaneous operation in two reference frames, the kicking player and the goalkeeper. Similarly, in a game of Rugby the attacking player is often in close physical contact with the defenders of the other team. While taking his opponent as a reference frame, he must also make an assessment of the positions of all his teammates, which requires a global view of the field. This ability to dynamically shift from one reference frame to another is probably one of the secrets of success in this kind of game.

This manipulation of spatial reference frames is a basic form of vicariance. Moreover, it is not limited to perception of space. To maintain your balance you have only to touch a stable point (a wall, the edge of a table, and so forth). This contact does not serve simply as an aid in allocating effort. The brain takes the point as a reference frame for organizing movements. This principle is actually used for humanoid robot movements by a robotics laboratory in Japan (the Joint Robotics Laboratory in Tsukuba) and is only one of the many features of the functioning of the human brain.

The Neural Basis of the Ability to Change Perspective

To understand the neural basis of this remarkable capacity to change frames of reference, our group has had to construct specific research paradigms that capture the passage from one reference frame to another, and the ability to use them simultaneously.

The first condition for using the repertoire of spatial reference frames relating to different modes of action and choosing one, or for handling a problem from different points of view simultaneously, is to have a single, coherent, and stable perception of one's own body and of its relationship to the environment. This construction requires multisensory integration, that is, a combination of the five canonical senses plus the vestibular sense and proprioception. In addition, the brain has to activate the body schema, which, as I explained in Chapter 2, is composed of simple or multiple internal models of the body, both in part and whole. We know that the temporoparietal cortex and the superior temporal sulcus—the perisylvian vestibular cortex—are areas of the brain that are essential for creating a unified perception of the body and its relationship to space.[3] This region of the brain has also been identified as important for social relations.

The paradox is that in parallel with uniting the body in a unique schema, the brain uses various coordinate systems of these different reference frames: in the retina, space is coded in the "retinotopic" coordinates; in the superior colliculus, space is also coded in this way, but with a modified geometry.[4] I suggest that different geometries are needed to plan trajectories in different spaces. You don't need the same geometry to pick up a rose lying on the table and to imagine your route through a city. In the parietal cortex, the spatial coding is egocentric, that is, it takes the body of the observer as the frame of reference. In the hippocampus, the spatial coding is allocentric, like a geographic map. On the other hand, in the putamen, the neurons only know the relative position of the limbs, although the coding can be activated by movement. Thus, the first challenge is that we do not know how the brain coordinates all these spaces and uses them to change perspective.

In addition, we know from neurology and recent brain imaging data that different neuronal networks are involved in processing these different spaces. We even activate different networks within very nearby space (less than a meter), that is, grasping space, and the space of the immediate environment (a few meters).[5] In other words, space is divided into zones that may correspond to the actions that can be performed there. The mathematician Daniel Bennequin and I have formulated the hypothesis that the brain uses different geometries in different spaces. These geometries probably include affine geometry and even some types of non-Euclidean geometries. We know, for example, that the response to danger depends on distance: fear can take several forms—freeze, flight, or fight—based on the nearness of the threat. This modularity poses challenges in coordinating the different spaces mentioned above, which is, in my opinion, a contributing factor in various pathologies, but also a powerful tool of vicariance. It allows humans to solve a problem by using several combinations of these various spaces and the neuronal networks that underlie them by calling into play "cognitive strategies."[6]

It is also possible to memorize an itinerary in the "mental palaces" described by Frances Yates and Mary Carruthers in their books on the art of memory.[7] These authors explain that, since the time of the Greeks, humans have used the mental palace as a technique for memorizing and storing ideas, objects, even verses of the Bible. In mentally wandering through these palaces, one can retrieve these ideas and even, as proposed by Mary Carruthers, construct new combinations of relationships. A priest could store elements of the Bible in different rooms of a mental palace and every Sunday combine them in various ways to produce a new discourse! Here, space is used for vicarious creation.

Cognitive Strategies for Navigation

A traveler strolling about a strange city, or a mountaineer making an ascent, will frequently plan their path by going over the route in their head. They can do this by taking advantage of diverse cognitive strategies. These strategies constitute a repertoire that can be called upon depending on the task, the context, and the person's gender, age, and culture. They correspond perfectly to the definition of vicarious functioning. In the words of Maurice Reuchlin, they are "evocable." They rely on brain mechanisms for processing space, or rather, spaces.[8] And they correspond to various vicarious ways of navigating around these spaces. They take account of the uncertainties of navigation as when, for instance, in a city, the cues we use to figure out where we are are ambiguous or our memory is unreliable.[9]

The first of these cognitive strategies might be called the *egocentric* itinerary strategy. When you visit a city on foot or by car, you employ this strategy to remember your movements, the detours you had to make, and to associate them with the visual landmarks you noticed and the events you experienced. We call this strategy "kinesthetic route memory." It is not limited to a simple association of movements and sensory data. As Jean-Luc Petit notes, kinesthesia is the essential operator of the perceptual constitution of a thing.[10] It enables you, as the perceiving subject, to give continuity, structural organization, and synthetic unity to the appearance of momentary sensory fields. Perceived things then emerge from these fields. But the role of kinesthesia should not be reduced to the solipsistic perception (by the solitary subject) of things, because kinesthesia is at the core of empathy. Indeed, it participates in intersubjectivity, as understanding, by analogically transferring into the body of another agent a portion of the intentional actions that the subject accomplishes, and the intentions that precede and accompany these actions.[11] In this way, the surrounding world is constructed by the brain based on successive "views" or sequentially organized points of view, encounters, and events that occurred during the stroll. This process is fundamentally egocentric in the sense that the perspective that is the basis for the analysis of the world is a "first-person" perspective. The imagined recall of the trip is a mental simulation of the sequence of the movements experienced. The flow of the experiences is reproduced, and the subject is, in a certain sense, a prisoner of those experiences. Even if the route is continuous, the brain tends to break it up into stages, which sometimes lends it the character of a cartoon strip but is helpful to memory. Thus, Hiroshige, the Japanese painter, illustrated the journey from Edo to Kyoto through a series of sixty prints. These prints divided the journey into very specific stages, in contrast to the many travelogues that attempt to retrace the progression through narratives.

The second strategy for remembering a trip has been identified by psychologists as a view of the whole, that is, an *allocentric,* or overview strategy. It allows you to construct a mental map of the environment, a map on which you can follow an itinerary as well as on a real map. Think of the area where you live and the path you take to get to the nearest bakery. You can imagine either the route (the first strategy) or the mental map of the neighborhood (the second strategy, called allocentric because it isn't centered on your body). The components of the environment are, in fact, linked to each other with no reference at all to the body of the person who is examining the space. The overview strategy, which consists of visualizing a map, is important when you are trying to remember long distances or to plan one route among several possible ones. It is probably the origin of our capacity to do geometry. It appears late in infancy, around the same time as structures like the hippocampus and the prefrontal cortex are developing.

The third strategy is called *heterocentric*. If somebody asks you how to get to the post office from their hotel, and you describe the path they should take from their point of view, you have to treat that person as the point of reference. This decentering also occurs when you are trying to understand the perspectives of the people involved in an argument.

A fourth strategy is one where, for example, you are attempting to remember the location of a colleague's office in a building. In this case, the brain creates a representation of the building thanks to a sort of transparent model of it. I call this strategy the *3D model*, for it involves building a mental model of a three-dimensional structure. Few studies have been devoted to this question. Yet it is useful when you are moving around the subway, in a big department store or a hospital, and so forth. Together with colleagues from the French electric corporation EDF, I have carried out two large studies on this subject having to do with the problem of getting around a power plant.[12] The first study focused on navigational strategies for emergency signage. We subsequently explored the differences caused by vertical or horizontal exploration on memorization of landmarks encountered in virtual buildings that schematically represent 3D architecture.

Other strategies have been proposed: "taxon" navigation (this is what insects do in moving toward a goal, and it is the most primitive form of navigation), and "choreographed" navigation, which refers to situations when movements are programmed—the path of a ballerina onstage is one example. What matters for our exploration of vicarious processes is that different networks of the brain are called into play depending on the strategy. Vicariance, that is, the capacity to replace one strategy by another, relies on switches in networks. Essential to our theory of vicariance is that the human brain possesses a repertoire of mechanisms that it can "play" with to construct alternative solutions when problems arise.

In everyday life, we continually change our perspective, and sometimes we use several viewpoints at the same time. I have proposed the idea that the choice of a slanted perspective in the presentation of navigational aids (GPS) is due to the fact that presenting a path from this perspective enables the brain to code it using both the egocentric and allocentric reference frames.[13] This enables the brain, on recall, to use the information memorized in one or the other of the reference frames based on the task and context.

How to Study These Strategies

We have developed experiments for studying cognitive strategies in situations of uncertainty. One of them uses a structure called the Magic Carpet, a collection of luminous blocks arranged on the ground that is derived from the Corsi block test, a spatial memory test that uses an array of blocks on a table. The tester taps on a series of blocks in sequence, and the subject must repeat the sequence. People with normal memories can usually remember up to seven locations. Cecilia Guariglia and Laura Piccardi in Rome have also designed a walking Corsi test. In our Magic Carpet, luminous tiles are placed in an array on the ground. A subgroup of them can be lit all together or in sequence. The subject in the experiment must remember which tiles were lit up and then step from one to the other in the same order. One can also change the perspective by showing the pattern from one side of the room and then asking the subject to reproduce the pattern starting from another side. This array has been used to study cognitive strategies in visuospatial memory in young children (with or without cerebral palsy) and elderly people (with or without mild cognitive disorders).[14]

I have also proposed a new hypothesis to our psychiatrist colleagues: *The manipulation of spatial reference frames is a "transnosographic feature" of various neurological and psychiatric pathologies.* By that I mean that each of these disorders (autism, schizophrenia, epilepsy, spatial anxiety, Parkinson's disease) affects some of the structures involved in manipulating the reference frames for different spaces. Consequently, each of these disorders shows symptoms of the deficits specific to each. Some cause perceptual and motor disturbances; others cause problems of orientation; others result in difficulty in changing perspective. These vicarious deficits lead to stereotypical behaviors or, as I will show later, difficulties in interacting with others.

Inter-individual and Inter-sex Differences

Many constraints prescribe precise rules of conduct on humans, for example, standards of teaching and learning; exacting industrial processes and procedures; judicial laws, particularly in France; multiple social codes; and the frequently monotonous cityscapes imposed on us by architects. The existence of these rules can give the impression that we are all subject to constraints in similar ways. But that is not the case. For example, researchers make a distinction between *assigned work* and *actual work*. Each brain is different, and each individual often takes a unique approach. This is the strength of the Anglo-Saxon model, which favors individuals and makes them accountable for both successes and failures, in contrast with those models, more familiar to Europeans, that emphasize conformity, stigmatize failure, and do not place trust in young people—who are, however, the best source of vicarious creativity—thus stifling innovation.

Consider the example of the diversity of individual competence in spatial processing and manipulating spatial reference frames discussed above. An important factor in the role that the treatment of space plays in vicarious processes is the difference in cognitive strategies between individuals of the same sex and of different sexes. People of the same sex have different perceptual styles, that is, they choose different sensory data.[15] Different individuals have different strategies: some preferentially use vision, while others favor "kinesthetic" information. Moreover, people use these cues differently depending on the context. The same person may employ a varied repertoire of perceptual styles, which constitutes a unique form of vicarious functioning.

Important differences also exist between the sexes, some of which have been summarized in works by psychologists Doreen Kimura and Melissa Hines.[16] For example, women depend more on vision for spatial orientation. They are statistically better at cognitive tasks that can be mediated verbally. And in a visual scene containing many objects, they are better at memorizing shapes that can be named and recognizing which ones have been subtly shifted or removed (the "seven errors" game). Men and women, or the anxious and the nonanxious, do not adapt in the same way to sensory conflict. Men are better at tasks requiring spatial mental rotation and maplike representations.[17]

These differences are not due to education, although obviously that can have a major influence. They derive from anatomical differences in the brain, but also from hormonal factors—variations in estrogen levels for women and testosterone for men.[18] For example, at different times in the menstrual cycle a woman will perform the same spatial test differently. In men, seasonal variations in testosterone levels are observed. Men with a low level of testosterone often perform better at spatial tasks.

In remembering routes, women tend to adopt strategies that are more egocentric and sequential. I believe this to be linked to their preference for verbal mediation: they prefer verbal descriptions of routes, which are by definition sequential. Later I will show that this preference is probably associated with brain lateralization. Men prefer allocentric strategies. Generally speaking, they are statistically better at spatial tasks requiring a mental rotation, as in changing perspective while reading a map. Our group has suggested that mental representations of large-scale environments by men contain more "metric" (cartographic) information than do those by women, the latter relying more on information about landmarks present in the environment during navigation. We have confirmed this observation in a supermarket.[19] We found that men also employ egocentric representations when they perform a finger-pointing task in the direction of a landmark. Such studies reveal genuine vicarious functioning, that is, a variety of strategies depending on sex, context, goal, learning, age, and occupation—compelling evidence for the brain's use of vicarious processes.

Social Consequences: Disorders of Spatial Vicariance

Differences between the sexes are also revealed by a different response to psychiatric pathologies and disorders. Spatial anxiety is more frequent in women; autism in boys; agoraphobia in girls; kinetosis (motion sickness) in women. Note, however, that when I write "women" or "men," I am painting with a broad brush and do not mean all women and all men. Moreover, people of both sexes are likely to use different strategies depending on education, learning, context, or objectives. It would be interesting to know whether these differences also affect how boys and girls learn mathematics and geometry, geography, and history.

Lateralization

How do the two hemispheres of the brain cooperate in formulating vicarious solutions? Do men and women use their "two brains" in the same way? Some cognitive functions are in fact lateralized.[20] The left brain, which contains the language center, is more concerned with details, and the right brain with global aspects of spatial shapes. I will not try here to summarize this question, which is very important to empathy, since each of the two brains plays a different role in emotion. Lateralization of the functions of the hippocampus appeared in primates, including humans, during the course of evolution. We have established that the right hippocampus specializes mainly in allocentric coding of spatial relations and events, whereas the left hippocampus specializes in sequential egocentric memory of routes traversed, and in episodic memory. This was shown by two studies that I will describe briefly.

In the first study, we proposed to patients with damage to the hippocampus that they take a little trip through a very simplified virtual village—a network of virtual hallways. Objects were situated at each intersection along the "route."[21] At each crossroads, the patient had to turn in his chair to switch hallways. The patient thus associated an object situated at the intersection with the rotation of his body to the left or right. Patients with damage to the left side could not remember the order in which they had encountered the objects (sequential egocentric memory). They also had difficulty associating with each object the rotational movement of their body necessary to navigate in the virtual world. Just as in the case of verbal description of itineraries, the sequential and egocentric mental simulation of the trajectory is done by the left brain.

Lateralization was confirmed using another task of navigating in a virtual world, navigation in a "starmaze," a maze created for mice by neurobiologist Laure Rondi-Reig that has five alleys radiating out from a central ring. We used the maze to study egocentric and allocentric navigation strategies and adapted a virtual reality version of it for humans.[22] Subjects trying to reach a goal in the maze could adopt either an egocentric or an allocentric strategy. We recorded their brain activity by functional magnetic resonance imagery. During allocentric performance, the posterior right hippocampus was activated; during egocentric performance, the left hippocampus was. This was a very good demonstration of the dissociation between the role of the right side in allocentric memory and that of the left side in sequential memory of movements, events, and landmarks.

We also established that the cerebellum is involved in online control of navigational paths.[23] The lateralization of hippocampal functions is duplicated by an interesting association with the cerebellum: the left hippocampus corresponds to the right cerebellum, and the right hippocampus to the left cerebellum.[24] This discovery established the close relationship between the cognitive aspect and the motor organization of human spatial memory. It is the lateral portions of the cerebellum that come into play. These areas, called the neocerebellum, are involved in pathways that link the cerebellum to the thalamus and to the cortex and then loop back to the cerebellum. Together they constitute the "cognitive" role of the cerebellum and are believed to play an important role in learning. As some of these pathways also involve the prefrontal cortex, they may participate in vicarious functioning. The future will tell.

Does Vicariance Govern Geometric Processing?

Conventional education convinces us that space is an abstract notion, independent of the "feeling space" of the moving body. This formalist view contradicts the ideas of great mathematicians. Henri Poincaré describes how, when a frog is decapitated and a drop of acid is placed on its skin, the animal attempts to wipe off the acid with its nearest foot. If this foot is amputated, the frog wipes the acid with the other foot. That is vicarious behavior. "There," writes

Poincaré, "we have the double parry . . . allowing the combating of an ill by a second remedy, if the first fails. And it is this multiplicity of parries, and the resulting coordination, which is space." Note the verb "is" in Poincaré's statement. For him, space means action space, and the vicarious duplication creates the space. In other words, the concept of vicariance is to be found even in the basic elements of geometry!

Poincaré's book *On the Foundations of Geometry* contains a veritable redemption of the body and of movement.[25] He hypothesizes that human geometry consists of associations between movements and their consequences. Thus, the foundations of geometry are to be sought in the most basic organization of movement. Poincaré writes in *Science and Method* that it is the "potentiality of warding off the same stroke which makes the unity of these different parries [as in the example of the frog], as it is the possibility of being parried in the same way which makes the unity of the strokes so different in kind, which may menace us from the same point of space. It is this double unity which makes the individuality of each point of space, and, in the notion of point, there is nothing else."[26] Geometry is founded on movements oriented toward points.

Here Poincaré goes a step further in his argument. He concludes that movement is basic to defining space because "a conscious being fixed to the ground" would not recognize space as such and would know it only as "changeless." Poincaré then goes into a long rumination on why he considers space to be three-dimensional. He distinguishes perceived ("representational" or "phenomenal") space from physical space. He then embarks on a quest for the true origin of the laws of geometry. His conclusion? That they were the result of human observations of the changing shapes of objects during movement. When we move, objects may appear to change shape for two reasons: either the object has moved with respect to us, or its state has changed. For Poincaré, the movement of our body is essential to the construction of a notion of space. An immobile being with no muscular sensations apart from sight would not be able to recognize these changes of position or state. Movement provides an opportunity to modify our interpretation of the properties of the world. This capability is a form of vicariance close to that of von Uexküll: we attribute to the real world properties that depend on our repertoire of planned acts and needs.

An Example of Mathematical Vicariance: The History of Imaginary Numbers

When I was in an engineering preparation class in Paris, we were asked to draw the intersection of a hyperboloid (a 3D form made by rotating a hyperbola around an axis) and a paraboloid (a similar form made by rotating a parabola). All my classmates could solve the problem by algebraic calculus. I could not figure out this method. However, I could solve the problem very quickly by a mental operation using "imaginary cyclic points." This is a beautiful example of vicariant cognitive strategies.

Not infrequently in mathematics, posing a problem gives rise to new ideas. The underlying process is akin to a form of vicariance. The mathematician Daniel Bennequin kindly provided the following examples. One is the concept of calculus, originally invented by mathematicians to calculate (and better understand) the tangents of curves, and the measurement of surfaces, using integrals. Another is the idea of imaginary numbers, introduced to solve whole-number equations (such as Fermat's conjecture). These new ideas themselves enabled the formulation of new problems and the development of new methods that make mathematics so deeply interesting. For example, imaginary numbers made it possible to extend the notion of the algebraic integer, to define different kinds of algebra, and to pose the question of the relationship between arithmetic and geometric functions (harmonic analysis).

Although new mathematical notions sometimes coincide with the solution to a problem, these notions themselves generate new problems, followed by new ideas or new mathematical objects. These creative mutations are characteristic of vicariance. The history of complex numbers offers a nice illustration. Since ancient times, mathematicians have attempted to introduce imaginary quantities as an aid in solving second-degree equations, by taking the square root of negative numbers. Heron of Alexandria is said to have been the first to have proposed this idea, and in the ninth century Mūsā al-Khwārizmī did so as well. However, these imaginary quantities were not scientifically recognized until the sixteenth century. The trigger appears to have been the attempt to solve a third-order equation with ordinary whole-number coefficients, having only one solution: Let $P(x) = 0$. The solution is very real, as can be seen by tracing the curve $y = P(x)$, which necessarily transects the axis at $y = 0$. The magic formula of the solution to the equation in question exists, and in certain cases it requires the use of imaginary numbers. The intermediate calculation is thus imaginary, but the final result is quite real. This is a simplex detour.

It took a "real solution" to a "real problem" to justify the new idea. The first systematic mathematical treatise on calculations involving imaginary numbers, or "complex numbers," was that of Rafeael Bombelli (1572), who showed how to apply the four mathematical operations to imaginary numbers. An initial form of vicariance thus appeared in using established mathematical rules for new entities. But there were other advances. In the centuries that followed, mathematicians such as Descartes, Leibniz, Newton, and Euler developed the mathematics of complex numbers (in fact, a simpler calculus: they could have been called "simplex numbers"), by extending it to include the emerging analysis of infinite series. These great thinkers passed on this imaginary tool, even while considering it an impossible but beautiful, powerful, and elegant thing. Indeed, the roots of negative numbers, employed to solve algebraic equations, made possible the demonstration of many remarkable entities.

Things suddenly changed when, at the end of the eighteenth century and the beginning of the nineteenth, Caspar Wessel, Jean-Robert

Argand, and Johann Carl Friedrich Gauss discovered that a complex number is nothing more than a point in a real Euclidean plane, with the addition and subtraction of vectors (the parallelogram rule) and with multiplication and division given by similitudes. Complex analysis owes its existence to this extraordinary coincidence of the representation of the imaginary number by a vector and by a similitude transformation. This is more about serendipity than about vicariance, but one might also say that Euclid's old plan was repurposed to extend the notion of numbers.

Thus, geometry came to the aid of algebra and analysis, and a host of new questions emerged regarding complex numbers. The arithmetic of Gaussian numbers, $a + ib$, where a and b are integers, held the secret of Pythagorean triangles. It became evident that they also contained the key to projective geometry and elliptical integrals. Consider also the advantages of "cyclic points," which are useful, for example, in determining the intersection of two complex surfaces (as in the example at the beginning of this section).

Then came Fourier transforms, and many other areas of the theory of partial derivative equations, such as one that was recently formulated by algebraic analysis, where complex numbers are more natural than real numbers. Finally: quantum mechanics, where the wave function that describes subatomic particles assumes imaginary values. Thus we see that imaginary numbers, initially invented to solve algebraic equations, have proved essential for solving many geometric, analytical, and physical problems: vicariance à la von Uexküll!

Changing Perspective: How Many Strategies?

The collection of mechanisms described thus far makes it possible for humans to change their perspective to find vicarious solutions to a problem. Suppose that you are sitting in a chair, reading this book. Now try to imagine the room you are in from another point of view. The brain is capable of performing a kind of mental displacement, an imagined rotation of the body in space. These manipulations of perspective require a mental rotation between the self and the objects in the environment, but also a process called "spatial updating."[27]

Here is an experiment done by our group that you can do yourself. We placed an object, a capital letter F, around one meter high, in the center of a room.[28] The subjects started at one point in the room, where they looked at the object with their eyes open. Then they were asked to close their eyes and, guided by the experimenter, to walk in the dark along a path formed by two perpendicular segments (like the two sides of a rectangle). When they reached the end of this path, they were facing the letter (still in the dark). The subjects had to carry out this task using two different cognitive strategies: first, keeping the letter in mind as they were moving, otherwise known as the continuous spatial updating strategy of displacement; second, forgetting the letter and concentrating on their own movements, and then, after reaching the goal, bringing to mind the shape of the letter they saw at the start of the exercise and mentally rotating it to perceive it from the new perspective.

There are (at least) two very different cognitive strategies for changing one's perspective and performing the spatial updating in this task. The first strategy allows one to have a good representation of the letter as soon as one arrives at the new position, whereas the second requires some time to perform the mental rotation of the memorized image using kinesthetic information linked to walking.

The Neural Basis of Changes in Perspective

The brain networks involved in changing perspective are becoming better known. We recently showed that patients with damage to the hippocampus have difficulty localizing an object in space and then changing their perspective.[29] In an experiment, we introduced an avatar into a virtual room. One task that involved localizing the object in the room required that the patients be able to mentally change their perspective. These patients had difficulty putting themselves in the place of the avatar, that is, in adopting the point of view of another person, even a virtual one. Other studies have shown that patients with damage to the thalamus and the pulvinar, a part of the brain involved in visual processing and gaze that projects into the cortex, were perfectly capable of locating the towns around Lake Geneva seen from above, as from a helicopter (allocentric perspective). However, they could not recognize the spatial relations between the towns of Geneva and Vevey from the vantage point of Lausanne: they were incapable of adopting an egocentric point of view of an imagined location.[30]

The participation of the hippocampus in changing perspective has also been suggested by clinical studies. It is known that manipulating reference frames involves the frontal oculomotor field. The retrosplenial cortex may also play an important role in changing reference frames and perspective, for it contains neurons that respond to the direction of the head in space and is activated in various alterations of perspective. We have found it to be activated in several spatial navigation experiments involving change in point of view.

Summarizing the work on this subject is beyond the scope of this book. Note, however, with respect to the neurophysiological basis of vicariance, that different navigational strategies, and the multiple spatial reference frames they rely on, call into play neuronal networks that are distinct but that ultimately employ the same structure (here, the retrosplenial cortex). Indeed, the principle of substitution, which is essential to the notion of vicariance, assumes that a structure like the retrosplenial cortex can be replaced by other circuits that provide the same information. To the best of my knowledge, no experiment has yet demonstrated this. The future will determine whether this hypothesis is valid.

Vicarious behavior is thus supported by a repertoire of functional models of the brain that specialize in each of these strategies. It is easy to imagine the wealth of possible combinations that this vicariance results in, so that we have the ability to imagine our actions in the worlds we create. It is also why I maintain that vicariance is a powerful tool for creativity.

Many Roads Lead to One Decision

To choose between several processes leading to the same objective—that is, to control vicariance—the brain needs to make decisions. Now, decision making sometimes means choosing among several solutions that ultimately lead to the same result. Decision making thus falls under the category of vicarious processes. I discussed this question in an earlier book (*Emotion and Reason*), particularly the need to rethink a biological theory of decision making that takes into account, among other things, the role of emotion.[31] Since that book appeared, much work has been done. For example, fundamental research has been carried out in monkeys on the neural basis of decision making and its association with the mechanisms of emotion. Other research concerned with the new field of neuroeconomics focuses on experiments involving financial decisions, opportunity loss theory, and counterfactual decisions. But it is not my intention to go deeply into this very extensive field.

What I would like to do is to very briefly describe the results obtained by the group directed by Étienne Koechlin in Paris.[32] This work is exemplary of a modern approach that is both theoretical and experimental and is very relevant to the daily activities of humans. Koechlin studied the role of the frontal and prefrontal cortex in the process of decision making. These structures have what is called an executive function: they coordinate and select behaviors appropriate to a task. They choose strategies depending on context and contribute to decision making.

In everyday life, we must adapt our behavior to situations whose outcome is uncertain, using information that is often incomplete. Here, vicariance refers to the brain's capacity to choose behaviors most likely to be effective with respect to the goals of the actor. Koechlin compared three explanations—three models—for understanding the role of the frontal and prefrontal structures: a probabilistic (Bayesian) explanation; another based on evaluating performance in a given situation; and a third that considers the effectiveness of past action. The subtlety is that actions are chosen and evaluated based on values that maximize reward (reinforcement learning), whereas strategies (or action plans) that regulate the choice of actions are selected for their suitability to the environment (reliability). This viewpoint is opposed to a purely neuroeconomic approach and introduces the notion of reliability and predictability in the selection of strategies (which would also explain aversion to risk).

Koechlin draws three interesting conclusions. The first is that in decision making, the human cortex cannot handle more than three or four different strategies. The second is that the model that best matches the data is based on the use of memory of past action. The third is that if the brain can determine that a strategy is likely to be reliable, it will choose that strategy to guide behavior. If, on the other hand, the brain cannot come up with a reliable strategy, it invents another one. This goes to the heart of the transition from simple vicarious functioning to vicarious creativity. Uncertainty is thus a powerful motivator for inducing the brain to search for novel solutions.

Part II

Ontogenesis and Plasticity

5

The Stages of Vicariance

How Vicariance Develops In Children

The various forms of vicariance appeared over the course of evolution. They become more exact, refined, and specific during the development of an infant. It is said that ontogeny recapitulates—or rather, *characterizes*—phylogeny.[1] Environmental and social enrichment has been shown to facilitate many brain functions and is certainly a powerful factor in the acquisition of vicariance. Conversely, children who are raised in a very poor environment may suffer deficits from the lack of stimulation and affective contacts. Also, the ability to substitute one motor or cognitive strategy for another is an integral part of the process of development. I will not linger over this topic here. I wish only to show the generality of these functional substitutions and to discern their mechanisms.

The Stages of Maturation

The maturation of cognitive function results from reorganization of the brain circuits during infancy.[2] The brain of an infant consists primarily of local networks. Subsequently a distributed connectivity is established by virtue of long-distance links that enable construction of functional networks involving various parts of the brain. Maturation of a specific region makes it possible to access a particular function (Broca's area for language, for example). Then circuits are created that link regions for a particular function. In this way, the circuits linking the thalamus, the basal ganglia, and the cortex enable choice of action. Finally, learning makes it possible to acquire new capacities, and each infant employs different networks to acquire the same capacity.

This latter mechanism is one of the fundamental bases of vicarious functioning. For example, the addition of the inferior parietal cortex to a network results in a new capacity for visuomotor tasks. On the other hand, damage to these new connections can induce a behavioral deficit. In other words, it is the link that is primarily involved in relearning, and not each structure. Moreover, it is necessary to understand the link that is established for each individual infant during learning. It is possible that these long-distance connections also contribute to the emergence of awareness.[3]

Psychologists have described the developmental stages in the relationship between an infant and its personal body, the world, and other people. Many classifications have been proposed. These classifications reflect reality to a certain extent, but they do not necessarily correspond to a fixed schedule. Significant temporal variability exists among infants. The challenge is to describe the neurobiological development that establishes substitution strategies, as in the examples that follow. Cognitive maturation in infants is a progressive process that offers ever greater opportunities for variation in strategies.[4] The replacement of one cognitive strategy by another during infancy—that is, vicarious functioning—occurs through redistribution of the networks involved in the same task.

Inhibition and Vicariance

Vicariance uses inhibition. For example, Olivier Houdé's group has designed a task that requires subjects to detect logical errors.[5] The task is a perceptual one, involving recognizing visual shapes and their logical relations. At first the subjects (both children and adults) use their parietal cortex and draw erroneous conclusions, because they process the problem mainly at the perceptual level. After training, adolescents and adults develop mental tools for logical processing that involve the prefrontal cortex, and they make fewer errors. This shift from the back to the front of the brain accompanies the appearance of a new capacity of inhibiting erroneous judgments to the benefit of correct logical reasoning.

In the area of spatial perception and reasoning, infants begin by confusing distance and number. If objects are arranged in a row on a table, a child will assess the number by the total length of the row: very few objects spread over a long distance will be judged as numerous. This is the "length = number" strategy. Olivier Houdé has shown that, during development, a child replaces this strategy with another, which consists of counting the objects, by inhibiting the initial length = number strategy. Here, the vicarious replacement is enabled by an act of inhibition. Elsewhere I have stressed the importance of inhibition and the fact that all the major centers of the brain involved in decision making and assuring transitions between movement control and thought are inhibitory. Inhibition is the greatest invention of evolution; it is also a powerful tool of vicariance. It allows redistribution of the networks involved in a particular function.[6] A special relationship has been shown between the cognitive capacities of children of preschool age and the structure of a specific brain area, the anterior cingulate cortex.[7]

Epigenesis and Critical Periods

The changing configuration of networks that accompanies and triggers functional maturation and enables the capacity to choose behaviors best suited to a situation is not a continuous process. We now better understand the existence of significant moments in development called *critical periods*. The notion of a critical period was used, for example, by ethologists to describe maternal recognition. I have discussed its importance elsewhere. In the context of the theory of simplexity, I suggested that the mechanism of critical periods makes it possible to avoid having to simultaneously organize all functions. This hierarchical spacing of implementation and maturation of specialized functions facilitates integration of skills and coordination of specialized faculties. From a constructivist perspective, this has the advantage of both rigor and flexibility in developing the ultimate dynamic structure of brain functioning. It is a principle of modularity and progressive assembly that is found in many activities, including industrial processes.

Here, within the framework of the biological theory of vicariance that I attempt to outline, critical periods are particularly advantageous. I suggest, though I cannot prove it, that critical periods are not useful solely for establishing isolated, specific skills in a very exact context at a specific moment in time. They also make it possible to situate each function in a context that gives it flexibility and enables it to contribute to numerous combinations of endogenous or external factors.

Now, vicarious functioning is impossible if each sensory or motor system, and each cognitive skill, is rigidly determined. The substitution of one function for another, and the combination of several elements of the behavioral repertoire to compensate for a deficit, demands that the components of this function be flexible. In the dialog about the role of nature and that of the environment, neurologist Takao Hensch and biologist Alain Prochiantz have contributed seminal discoveries concerning the visual cortex. They not only showed that activity and experience play a crucial role (this was established for vision by famous experiments such as those of Richard Held and Alan Hein in the 1950s); they also described the detailed neural basis of the role of activity on complex synaptic and molecular mechanisms and the expression of genes during the critical periods, as well as the essential role of inhibition in these mechanisms.[8] Among other discoveries, they identified the role that the transfer, under the influence of experience, of a homeobox protein (Otx2) plays in the plasticity of the visual cortex. I will venture to hypothesize that the control of activity during critical periods is not only stabilizing, and does not only facilitate the emergence of a function, but that it is essential for constructing a living organism that is adaptable, flexible, and *creative*.

Emergence of Vicariance

For an infant to be able to substitute one strategy for another, he must acquire the capacity to inhibit the automatic strategies of primary infancy: he must be able to examine solutions outside of his own perspective and to change his opinion. What follows are several examples of research by psychologists who have studied the development of cognitive function.

Philippe Rochat, for example, has distinguished six levels of self-awareness that are built up during the first five years of life:

—*confusion*: the infant cannot distinguish his mirror image from the environment.

—*differentiation*: the infant perceives a relationship between his body and his mirror image, which he differentiates from the external world.

—*situation* (around two months): the infant perceives his mirror image as clearly being linked to his own self and distinct from the environment, a "proto-self" situated in the world.

—*identification* (around two years): the infant clearly perceives that the image in the mirror is of himself.

—*permanence*: the infant perceives his image, even in videos, and acquires a sense of his own permanence.

—*self-awareness* and *meta-awareness* (around four to five years): the infant has the ability to project himself into others and to perceive himself in the third person.[9]

An important idea of Rochat is that, in adults, the play of identity exchange with others still involves these basic levels. If that is the case, they certainly play a role in "social vicariance."

Other roughly analogous models have been proposed that outline the development of self-awareness and of relations with the external world. One of these distinguishes five stages.[10] Between the ages of one and a half and three years, the child can envisage past or future events in relation to the present. For example, a child of two years could think that, from the perspective of the current moment, he went to the zoo yesterday to see the animals. This provides continuity to lived time. Between three and four years, the child can adopt a decentered perspective to analyze two events that happened to him at different times.[11] However, the history of the world and his own history remain confused. At four to five years, a child can link a past event to the present.[12] He can also decenter himself and distinguish two series of events, one experienced by himself and the other occurring in the world, and thus prepare the process of vicariance.

At the neuronal level, the pivotal turning point of seven to eight years—the beginning of vicarious play—is linked to maturation of the prefrontal cortex and its connection to the hippocampus, the parahippocampus (especially important in manipulating scenes),[13] the parietal cortex, the retrosplenial cortex, and all the networks involved in manipulating perspective and thus executive function.

"Executive function" refers to the brain's capacity for making decisions, comparing challenges, planning a series of actions, evaluating the advantages and disadvantages of a strategy or movement, changing point of view and opinion, and so forth. This capacity is attributed to the prefrontal cortex, but in reality, it works within networks that involve the subcortical areas. It is only from about four to five years that children can link a past event to the present.[14]

We have shown that a child's capacity to transition from an egocentric perspective to an allocentric one occurs between seven and ten years of age, although the allocentric perspective is present in a latent fashion in younger children (around five years).[15]

This maturation continues until puberty (fourteen to fifteen years and beyond), during which dramatic changes due to hormonal factors add motivation, mood, and emotion into relationships with others. These networks deteriorate over time and are less effective in the elderly. The capacity for vicariance diminishes with age. I will return to this subject later on, for, in the elderly, mechanisms of vicarious compensation appear (fortunately!).

Restoration and Maturation of Vicarious Functioning

On the whole, and despite their differences, psychologists describe more or less the same developmental pathway. Vicarious functioning or vicarious usage begins between five to seven and ten years of age. The psychiatrist and neuropsychologist Julian de Ajuriaguerra (in whose honor I, in collaboration with Fabien Joly, sponsored a symposium at the Collège de France)[16] described at length the diversity of factors that are involved, to which today should be added epigenetic mechanisms. I previously proposed four main hypotheses, inspired by the idea of vicariance, for rehabilitation, or remediation, of damaged functions in children.[17] These hypotheses could apply just as well to adults:

—Do not necessarily attempt to repair the damaged system; rather, aid the brain in developing another combination of networks to accomplish the same task (vicarious functioning).

—Retrain the body schema, that is, the mental representation of the body (its double), and not just the physical body. Retraining concerns not only muscles but also internal models. Study damaged reference frames and work to restore to the patient the capacity for manipulating body and spatial systems of reference.

—Retrain anticipatory and inhibitory capacities, including mental imaging.

—Retrain hand and gait movements simultaneously, or retrain one by means of the other. There actually exist shared simplex principles for the two types of limbs.

Mental Isolation Destroys Vicariance

What happens when a child is prevented from "decentering," from manipulating points of view, from acknowledging that there are several ways of acting to attain the same goal, and several ways of seeing the world and appreciating others—what I call social vicariance? The more specific problem is the age-old one of exploiting children to indoctrinate them, locking them into rigid mental schemas that promote intolerance, hatred, fanaticism, or mental dependence. It is important to state that children that are victims of this indoctrination are chosen at particular ages, between seven to eight and fourteen to fifteen years. This age range is typical of today's child soldiers, who are asked to commit murder. The indoctrination of children as apprentices in religious or political fanaticism is not new: the Hitler Youth consisted of children recruited and indoctrinated at a very young age. They were thrown into the Battle of Berlin. Pol Pot also made a practice of indoctrinating very young adolescents. Today, young adolescents are being indoctrinated to commit suicide attacks, and some cults practice brainwashing by isolating their followers from society.

I have proposed that the period between seven and twelve years of age is a critical period for acquiring the capacity to change one's point of view and therefore develop tolerance. The problem is that the existence of this critical period, although it simplifies relations with the world, also allows for the development of a rigid characterization of others if during that time the child is prevented from using new functions to acquire the possibility of many points of view or opinions of others. If a child is indoctrinated by others' interpretation of sectarian schemas, the child will tend to prefer these interpretations and will become locked into them. I have developed this theory in several publications.[18]

Vicariance is fundamental to having opinions about others and the world that can change. What are the mechanisms that enable mental isolation in children? It seems to me that, to fight against confinement within a mental schema and to promote tolerance, one has to be able to change one's perspective of the world and to manipulate representations and ideas.[19] Thus, what is needed is a capacity for vicariance, not just in the sense of replacement but also in the sense of being able to put oneself in another's shoes—to have empathy.

There are indeed critical periods, that is, moments of development or narrow temporal windows during which the potential for specific functions is activated. Ethologists like Konrad Lorenz have described a critical period for bonding with the mother. My hypothesis is that there is a critical period for the capacity to entertain several interpretations, and if a child is not allowed to exercise this faculty, his opinions will be rigid. If, at that moment, a child is locked within a narrow interpretative schema, he will remain "blocked" and will not be able to access his capacity for tolerance. This critical period is fundamental to the development of mechanisms that enable social vicariance.

6

Vicariance and Brain Plasticity

Suppose that a part of the brain with a specific function—language, planning movement, memory—has been damaged. Often, an attempt is made to re-establish the function through training, hoping that the affected area can regain its activity. In reality, this is often not possible. The solution is to allow the brain to substitute another area for the defective one, if possible, or to create a new combination of networks that can perform the same function. If your hand is paralyzed, you can paint with your mouth. And the other senses can substitute for vision in perceiving space. Today, we prefer the term "remediation" to "rehabilitation." At issue is vicarious functioning, which makes use of the brain's plasticity and flexible, creative capability. In the strictest sense, plasticity refers to the capacity to alter the functioning of the brain at several levels—at the level of molecular mechanisms, but also that of functional connectivity. However this is not the place to go into an exhaustive description of the remarkable capacities of brain plasticity. In the sections that follow, I will simply provide a few examples which I believe are relevant to the concept of vicariance.

Do Adults Also Experience Critical Periods?

I explained in Chapter 5 the importance of critical periods during ontogeny for the capacity of vicariance. Sensory or motor critical periods in development are moments when plasticity is expressed in infants and children. But adults, too, experience critical periods. Neurons that mature during adulthood can exhibit plastic properties like those that are characteristic of critical periods. For example, a new critical period for the visual cortex has been discovered in the adult brain. It does not manifest itself through the establishment of a function, as during ontogeny, but rather by the capacity to reveal a hidden function. For example, when a healthy rat is exposed to light, the neurons of the visual cortex normally respond by firing (action potential). But there is a subliminal—that is, hidden— response. Following damage that suppresses the normal response, manipulation of the balance of excitatory, or inhibitory and excitatory, influences, or suppression by the researcher of molecular "brakes" that affect plasticity, this hidden potential is "revealed" and is able to facilitate restoration of function.[1] We can call it vicariance through the unmasking of a latent property.

Sometimes, the plasticity of substitution is limited by molecular mechanisms involving the neuroreceptors that govern both the plasticity and the stability of already established circuits. One can then tinker with the regulation of synaptic expression to remove this limitation. For example, one-month-old neurons of the dentate nucleus in an adult animal can exhibit long-term potentiation and variation in thresholds, which indicates a capacity for plasticity.

The possibility of inducing functional plasticity in adult neuronal networks is promising for restoration of function. There is a report in the literature of a case of precocious reorganization of visual pathways in an infant born with only one cerebral hemisphere: the capacity of the brain to reorganize is remarkable.[2]

Neuronal plasticity depends on synaptic modifications linked to the temporal distribution of pre- and postsynaptic action potentials (spike-timing-dependent plasticity). For example, the very precise (paired) temporal association of one sensory stimulus with another alters the properties of the sensory neurons. In the primary auditory cortex of animals, stimulus in the form of paired sounds causes changes in the sensitivity of neurons to sounds. This plasticity depends on precise timing of sensory inputs on the order of milliseconds. This type of plasticity is fundamental for the shaping of neural function in the adult auditory system.[3]

The value or relevance of a stimulus can influence the integration of new information about the world and the plasticity of the circuits involved. For example, suppose that a rat likes to hear sounds of a particular frequency because they are very significant for its needs. In the auditory cortex of the rat, the paired association of the cholinergic stimulation of the basal nucleus, which is sensitive to the relevance of a stimulus for the rat, with a new auditory stimulus that is normally neutral for the animal, can produce a change in frequency of the preferred stimuli by the cortical neurons. This association causes a rapid reduction in synaptic inhibition within a few seconds, followed by a substantial increase in excitation specific to the associated stimuli. This period of disinhibition may be a fundamental mechanism of plasticity and may serve as a memory cue for stimuli or for episodes that have acquired behavioral significance for the animal.[4]

Experience and Plasticity

At the beginning of this book I emphasized the importance of action for vicariance. Here, I would like to touch briefly on the mechanisms that enable activity to participate in and contribute to establishing vicarious functioning in the brain.

Many factors influence brain plasticity: the balance of excitation and inhibition, protein kinases, transcription factors, chromatin, acetylcholine, noradrenaline, dopamine, NMDA receptors (glutamate receptors that detect coincidences between pre- and postsynaptic activity). Motivation, attention, interest, pleasure, and experience are essential.

In Chapter 5, I mentioned that recent work has established the importance of an enriched environment for all these factors.[5] What we mean by an "enriched environment" for a laboratory animal raised in a cage is one that contains many objects, games, or obstacles that provide the animal with the kind of environment that promotes the development of perceptual, motor, and cognitive skills. Such an environment increases cortical thickness and the density and complexity of dendritic spines, as well as plasticity. Hippocampal neurogenesis is also increased in older animals by enrichment. It is thought that these factors taken together may induce various restorative processes, that is, vicarious functioning. But the precise mechanisms are under study.

Indeed, experience alone can modify the anatomy and functioning of the brain independent of critical periods. It is known, for example, that exercise changes the sensory and motor brain maps (in musicians, amputees, and so forth). The training of working memory results in an increase in the density of dopamine D1 receptors. All it takes is ten hours of training over five weeks to produce alterations in the capacity of the prefrontal and parietal cortical neurons to bind dopamine D1.[6] The psychologist Eleanor Maguire has shown that in taxi drivers, the practice of driving in a city like London, which requires mental manipulation of maps for navigating, increases the volume of certain parts of the posterior hippocampus and decreases the volume of the anterior hippocampus. She also showed that expertise diminishes certain functions linked to the memory of objects in space.[7] Know-how comes at a cost!

Brain modification was shown to be associated with vicarious functioning in a study on a monkey's ability to grab an object with its fingers. This learning increased the cortical representation of the fingers involved in the task in an adult monkey.[8] What is interesting is that the monkey first tries various ways (as indeed a human would) of combining its fingers differently to reach the object. Once the most effective strategy has been discovered, it is very rapidly stabilized and the movements become stereotyped, which results in changes to the brain's anatomy. Vicariance is essential to determining the optimal coordination for each animal.

Another form of vicarious functioning in an adaptive process is, in fact, the rapid adoption by one of the cerebral lobes of a function that is damaged in the other. For instance, a model of rapid functional recuperation after a cerebrovascular attack can be obtained by stimulating the cortex with a magnetic field (magnetic transcranial stimulation), which blocks the activity of the stimulated cortex. This stimulation was used to probe the functioning of the left premotor cortex involved in selecting action. A rapid increase in activity was noted in the right, contralateral cortex, which suggested that it was taking over the function of selecting. The increase is linked to the moment when the action is being prepared, not when it is being performed.[9] This very specific functional reorganization allows for preserving a behavior and achieving a task following interruption of normal neuronal functioning—clearly a case of vicarious functioning at work.

Many studies have been done on brain reorganization following functional damage.[10] To end this discussion of the neural basis of vicariance, here is a spectacular example of structural change within the brain to compensate for a defect. A four-year-old girl underwent surgery on her occipitotemporal cortex to remove a brain tumor associated with Sturge-Weber syndrome. She had lost the capacity to visually analyze letters. Around the age of eleven, after she had learned to speak but before she could read, her brain activity was recorded by functional magnetic resonance imaging. The examination revealed that the visual area related to reading in the left cortex had moved to the right cortex, specifically, toward the visual area involved in analyzing forms. This displacement did not affect activities related to language, which remained localized on the left side.

The Bottleneck Hypothesis

A paradoxical consequence of cerebral vicariance may be bottlenecks in certain areas. For example, damage to the language region on the left side may be compensated for by activity on the right side. Yet the right brain is normally involved in visuospatial tasks. Thus competition may arise (whence the term "bottleneck") in the right hemisphere. This problem has also been studied in healthy subjects. In fact, in 1 percent of right-handed people and 10 percent of left-handed people, the language area is on the right side of the brain. Curiously, this shift does not always negatively affect visuospatial tasks. The inverse may be observed: that is, the left cortex may take over the visuospatial functions normally carried out by the right cortex. The brain does indeed possess many strategies for implementing vicariance.

How Mental and Motor Imagery Influence the Brain

In the previous sections, I discussed vicarious functioning resulting from interactions with the environment. A recent significant discovery is that mental and motor imagery can also enable the brain to find alternative solutions, in a manner similar to experience. For example, neuronal activity has been recorded in patients with Parkinson's disease during surgical procedures to lessen their symptoms.[11] Imagining a simple repetitive movement produced a change in the firing pattern of neurons in the internal globus pallidus, a structure known to be involved in motor control and selection. This is consistent with the general theory I elaborated in *The Brain's Sense of Movement* about the brain as simulator. It also shows that it is not only mirror neurons that are involved in imagining and simulating movement.

The Difference between Rehabilitation and Remediation

The last several years have seen a trend toward qualifying, or replacing, the term "rehabilitation" with that of "remediation." The latter term is also used in the field of education as well as in the area of soil pollution. For our purposes, that is, cognitive and sensory-motor functions, relations with others and social life, this dual terminology reflects different approaches to restoring function following damage, or when neurological or psychiatric illness or disorder results in major functional deficits.[12]

If you wanted to summarize the difference between these two methods in a very superficial way, you could say that *rehabilitation* consists of trying to restore the malfunction by repairing or activating the defective neuronal networks or sensory-motor system, whereas *remediation* consists of placing the subjects in a context of activity that enables them to replace the defective system by another one corresponding to their own capacities, as needed. Moshé Feldenkrais had the intuition that mental simulation of movement played a major role in retraining to transfer learning from one side of the body to the other.[13] This goes to the heart of the notion of vicariance as described in this book. It is not about copying a task— here the "vicar" must invent a new sermon! —but of accomplishing it without violating the general laws of nervous system functioning and the subject's private, social, or cultural context. I stress this last point because it is not a question of the brain finding a solution that contradicts, for example, the moral rules of the society in which the subject lives.

Psychiatric illness is commonly thought to manifest itself through various functional deficits. For example, autistic children are believed to possess disordered verbal communication. The disorder is even used to characterize the illness. Taking the opposite view, psychiatrist Laurent Mottron suggests that such inability to master language is actually associated with the freeing up of other skills, which in the healthy child are masked by the dominance of language.[14] The psychiatrist Nicolas Georgieff has proposed that autistic children and, more generally, people with psychiatric disorders, find new solutions to compensate for deficits caused by genetic mutations.[15] And the psychiatrists Nicolas Franck and Isabelle Amado have organized a "cognitive remediation" network intended to help people with psychiatric disorders find compensating solutions adapted to their pathologies.

Compensating for Age-Related Deficits

Vicarious processes also occur during the aging of cognitive and motor functions. In addition to the normal deterioration of motor function, profound changes in gene expression in the cerebral cortex occur between fifty and sixty years of age, suggesting that this age range is critical for the transition toward brain aging. Yet, paradoxically, in humans, other primates, and rodents, the loss of neurons in the hippocampus and the neocortex is not linked in any clear way to the normal cognitive deficits observed at this age. The human brain loses less than 10 percent of its neurons between twenty and ninety years of age. It is, rather, discrete changes in dendritic branching and in the density of dendritic spines, which are important in synaptic transmission, that are most characteristic of aging.

The interesting observation relating to the theme of vicariance is that the aging brain can preserve a relatively high level of performance despite alterations due to age. This preservation is the result of the involvement of new brain regions in a given task through the finding of new strategies to achieve the same goal. This is exactly what I mean by vicarious functioning. For example, a study comparing the performance of young people and elderly adults in a visual memory task showed that young people are more affected by transcranial stimulation. Researchers stimulated the dorsolateral prefrontal cortex during encoding, memory recall, and evaluation. The data suggest that the young people exclusively use areas of the right cortex for this task and show substandard performance if it is disturbed through probing, whereas elderly adults use the left cortex as a replacement for the right cortex in these conditions. The elderly thus produce a vicariance by exploiting the brain's bilaterality. There are many other similar instances of a function being assumed by a side of the brain that is not normally involved.

Another example is the increase in activity of the frontal cortex that is often observed in elderly subjects. The theory known as scaffolding proposes that this increase in activity is due to recruitment of other regions of the brain to compensate for defective structures that have become nonfunctional. The authors emphasize that this process is common in young subjects but that it also occurs in the elderly.

A final example concerns verbal memory. The dorsal part of the left inferior frontal cortex and the right posterolateral temporoparietal cortex are important in storing and repeating verbal information. An experiment with elderly subjects who showed improved performance revealed an increase in activity in these subjects for all categories of statements, whereas young subjects exhibited increased activity only in certain categories.

Part III

Vicariance and Sharing Emotions

7

Sympathy and Empathy

The example of forcing children into rigid patterns of interactions with others, which was discussed in Chapter 5, leads naturally to the question of how vicariance comes into play in social interactions. I have mentioned "social vicariance" several times thus far in this book. In fact, the twenty-first century has been marked by paradoxes that concern some of the most important aspects of social life. We are witnessing the expansion of both freedom and authoritarianism—for example, we have unprecedented means of expressing ourselves on the Web, but at the same time we face a drastic increase in fanaticism, especially evident in the case of children. The generous ideas of the Universal Declaration of Human Rights and the Enlightenment are being threatened by a stubborn resurgence of hatred of others. However, to defend freedom (a fundamental prerequisite of tolerance), people need to be able to change their opinions and perspective.[1] Sociology and social psychology have accumulated vast amounts of knowledge on this subject, but we must go further and try to understand the neural basis of social behavior. Work in this interdisciplinary area, known as "social neuroscience," has already begun.[2] I will limit myself here to saying that emotions are underpinned by numerous specialized networks (for the face, the body, and so forth). Once again we are dealing with the simplex principle of modularity, which is well documented in recent literature. I will cite only a meta-analysis that is relevant to vicariance insofar as it seeks to identify the different networks involved in emotions and thus avoids an analysis that would view emotions too narrowly, as governed by isolated areas of the brain.[3]

Consider one particular case, that of cooperation between two people to accomplish a common task. Here, the vicar is not replaced; he is present, a partner to share tasks with. A few years ago, Natalie Sebanz, Günther Knoblich, and Wolfgang Prinz reported work on this subject.[4] Here, vicariance is defined in a profoundly intersubjective way. The adjective "vicarious" is used, for example, when I feel the experience of another person simply by observing it, or even when someone recounts an event that happened to someone else. It is also used to describe the experience of a reader who feels the emotions and actions of the characters in a novel. In everyday life, we use other people, just as others use us, to help guide our actions. At a traffic light, we often wait for the other drivers to start before starting ourselves.

The use of others in order to act involves a subtle form of anticipation.[5] In the case of a task shared between two operators, neither constructs a common representation of the task to be performed ("What do we do?"), but rather, each one is concerned with identifying the contribution of others in terms of dividing up the time ("When is it my turn?").[6]

Vicariance and Emotions

In fact, it is impossible to construct a modern theory of vicariance without incorporating emotions into it. But I propose going a step further. I would assert that emotion is a powerful agent of vicariance, in the sense in which I would like to develop this concept, that is, emotion as source of diversity and creativity. Moreover, to safeguard the notion of the unity of humankind, we must not forget that the brain cannot function without emotions—an idea I developed in *Emotion and Reason*. Although Merleau-Ponty acknowledged the existence of modules, he urged us to recognize that the global balance of the organism includes the emotions. He went so far as to assume that, at core, each physiological variable (if you think of it from that point of view) reflects not only the specific function of which it is a sign, but also all of the others, and even a form of the "essence of the individual." In also highlighting vegetative variables (respiration, cardiovascular system, transpiration, and so forth), his ideas are in accordance with the more recent theories of Antonio Damasio on emotions. He incorporates the ideas of William James on the importance of vegetative functions not only as correlates of emotions but as "somatic markers" of the decision to act. Étienne Koechlin took into account the role of motivation in decision making. He proposed a three-part division of labor: the mediofrontal cortex is involved in *motivation* (I want an apple), the orbitofrontal cortex in *attribution of value* (I like apples), and the prefrontal cortex in *decision making* (I am going to eat the apple). He emphasizes that the problem of decision making is not reducible simply to a choice based on values but that it is also based on rules, adherence to which is not a choice based on values (I am following the rule because, in short, I prefer the consequences of following it to the consequences of not following it). The rule intervenes in the process of decision making by limiting as much as possible the influence of individual preferences on choice.

A Spatial Theory of the Difference between Sympathy and Empathy

Relationships with others involve at least three different processes, which are similar to those mentioned in Chapter 4 for the processing of space. First is the simple duplication of our person (the replacement by the vicar). It remains *egocentric;* it is an imitation, a resonance in the mirror, but without a change of perspective (as when a lover says, "I love you"). Next, there is the replacement of ourselves by another person as the origin of perspective, by imagining ourself in that person's place (when the person asks, "Do you love me?"). This is a *heterocentric* relationship. Finally, there is a relation independent of both of our perspectives, called *allocentric* (when a third person says, "They are in love"). The capacity to change perspectives in a relationship is a good example of the vicarious process.

In the sections that follow, I will examine two of the main vicarious behaviors that use a changing perspective and manipulation of spatial reference frames in personal relations: *sympathy* and *empathy*.[7]

Sympathy and empathy are vicarious behaviors that involve the sharing of emotions. We now have good knowledge of the networks involved in the production and perception of emotions and the influence of emotion on motor behaviors such as walking.[8] Recently, Beatrice de Gelder, a cognitive neuroscientist, suggested that several subsystems are involved in perception of the bodily expression of emotion. There is nevertheless an essential difference between sympathy and empathy. This difference was noted by the psychologist Carl Rogers, who wrote: "The difference between empathy and sympathy is important but not easy to describe. These feelings are related in that they represent resonance with the feelings of others. However, because sympathy pertains essentially to emotions, its perspective is narrower than that of empathy, which refers to both cognitive and emotional aspects of the experience of others. Moreover, in the case of sympathy, the participation of the subject in the emotions of others proceeds in terms of the subject's own experience of himself. . . . In the case of empathy, the individual forces himself to participate in the experience of others . . . based on the vantage of the person who is having the experience—not on the subjective point of view. . . . Both represent subjective forms of knowledge. But in the case of empathy, it is the subjectivity of others . . . that is at issue."[9]

Sympathy is close to emotional contagion, a resonant process engendered by purely automatic and very powerful mechanisms similar to those that control a baby's smiling when its mother smiles. A recent study on dyads showed that the expression of emotion in one of the two people triggers an emotional contagion, testifying to the power of sympathetic resonance.[10] In this case, we are in the throes of "sym-pathy" (from Greek *sum-patheia*, whose German equivalent, *Mitfühlung*, meaning "feeling with," was also used in German phenomenology). Thus, we attribute to ourselves what we observe in the other. This process of attribution may require us to localize or simulate the other in ourselves and identify with him. But in this process of sympathy, we remain in our own place and see the other person from our own perspective.

The capacity for sympathy is certainly due, in part, to the functioning of the "emotional brain" (amygdala, orbitofrontal cortex, and so forth) and to the system of mirror neurons, one of the great discoveries of contemporary neuroscience. In the brains of monkeys, mirror neurons are situated in the premotor ventral cortex (F5) and the inferior parietal lobe. These neurons are activated both when the monkey performs an action, such as grabbing a peanut, and when it observes the same action being performed by another individual or by the experimenter.[11] Actually, what is at work is a network of brain regions that also involves the superior temporal sulcus, which is responsible for the ability to simulate the actions of others. Brain imaging studies using functional magnetic resonance have shown the existence of a similar system in humans.[12]

This is a specific, rather simple, vicarious mechanism. It is not a matter of others as a replacement for me, but others "as myself." Several networks enable us to directly feel the actions, emotions, or intentions of others. For example, observing a facial expression of disgust bilaterally activates a region of the brain that is also activated by exposure to offensive odors. Having a brush stroke your leg, or watching someone else have their leg stroked, bilaterally activates the secondary somatosensory cortex in both cases.[13] In the frontal cortex, groups of neurons can detect a repertoire of expressive gestures (for example, aggression). This "embodied simulation" attests to the role of close links between perception and action in relations with others. In other words, the internal models that govern and correct one individual's movements can also be employed to understand the actions of another individual.

Sympathy entails imitation of behaviors. Humans are not normally aware that they are imitating. If your superiors treat you badly, you tend to imitate them in an attempt to form a bond. Imitating others also leads, in a largely unconscious fashion, to more obliging and generous behavior. One study focused on tips in a restaurant: tips were larger when the server (in connivance with the experimenter) imitated the customer by copying, for example, the customer's posture, gestures, or smile.

Another type of experiment has shown the dynamic aspect of perception that involves a process you might call "synchrony between two brains." For example, one electroencephalographic study recorded two people watching a film (a Western) at the same time. The study showed that the brains of the two spectators "resonated" very selectively.[14] In such situations, the fusiform gyrus, a brain area involved in the recognition and visual memory of shapes (faces, letters, and so forth) is activated synchronously in the two people. If you will permit me a speculation completely devoid of any proof, that phenomenon suggests a fundamental mechanism of the crowd effect. If a skilled and charismatic orator can synchronize the brains of a crowd, he can, through an effect of "sympathy," give the crowd the impression that he is representing their desires. He becomes their vicar, and the crowd is thus ready to delegate its power to him and accept him as leader.

Empathy

A second way of experiencing the pain, or joy, of others is to adopt their perspective and to feel their pain in their place. To enter into the body of another person, you must see the world from their point of view, that is, really adopt their perspective. This process was first described (and presented in this way) by the German phenomenologists at the end of the nineteenth century and the beginning of the twentieth century through use of the term *Einfühlung,* which literally means "feeling into."[15] My hypothesis is that in *Einfühlung,* the prefix *ein,* which means "in," has to do with a spatial process of mental simulation by which people project themselves into others. The root *fühlen* refers to a sensory or subjective emotional experience. Consequently, analysis of the terms *Ein-fühlung* and *Mit-fühlung* suggests that these two processes differ in the spatial and temporal significance of their prefixes. Empathy thus depends on mechanisms that are much more complex than those relating to sympathy. Empathy demands a change of perspective, and a certain form of "out-of-body" experience. It requires mentally separating oneself from one's own body and traveling in the other body with one's "mental double," the *Doppelgänger* of German literature.[16]

The neural basis of these changes in perspective has been studied in depth by Jean Decety and his colleagues, who attribute a fundamental role to the temporoparietal cortex and to the perisylvian cortex.[17] This is the region that Wilder Penfield indicated as being involved in "the awareness of the body and spatial relationships," and where my group has identified the "vestibular cortex."[18] At least four processes are required for one to consider that what is at work in an intersubjective relationship is a form of empathy.

1. First, one has to construct a unified and coherent perception of one's body and its relationship with the spatial environment. Deficits of empathy (and sympathy) are sometimes

actually defects in the construction of this coherent identity (body schema, body image, notion of self). Now, as I showed earlier, the construction of the personal body is the basis of the capacity for interacting with others. Maurice Merleau-Ponty wrote: "Husserl wanted to say . . . *that there is no constituting of a mind for a mind but of a man for a man.* By the effect of a singular eloquence of the visible body, *Einfühlung* goes from the body to the mind."[19]

2. Subsequently, as in sympathy, one must "resonate" with the emotions and intentions of others. I assume that, in the empathic process, in addition to being able to put oneself in the place of another, one must conserve the capacity to resonate with him as in sympathy.

3. One must also be able to change one's own perspective, and to mentally enter the body and brain of others (*Einfühlung*). This requires the intervention of one's mental double. Today, this notion of a mental double seems a little parapsychological, but it is supported by solid data that builds on the ideas of body schema or "body image" that have long been employed by neurologists. These ideas correspond to processes similar to those found in certain forms of pathology experienced by people with epilepsy or schizophrenia under very specific conditions, that is, "out of body experiences," "heautoscopy," and "autoscopic hallucination." These processes fall into the category known as "translocat[ion] of egocentric point of reference."[20]

4. The fourth, and final, fundamental component of empathy is that, to help others or to behave in a prosocial way, one must be capable of inhibiting the emotion one has felt in the place of others, and thus of suppressing emotional contagion. The doctor who wants to help a suffering patient must be able to remove himself from the intersubjective relationship and act in a way that enables caring for the patient. This aspect of

empathy is not traditionally taken into account. However, it is essential for understanding prosocial behavior.

In short, empathy requires one to be oneself and someone else at the same time. The transition from the first person to the third person involves both cooperation and competition between several brain networks and those of others, in a dynamic interaction.[21] Empathy is thus a fundamental element of social vicariance.

One idea worth delving into more deeply in this context is compassion. Notable here is the recent work of the group of psychologist Tania Singer on the connection between compassion and what she calls "empathic resonance," which I tend to call sympathy.[22]

The Virtual Tightrope Walker

Psychologists have studied how individuals act toward each other, especially changes in viewpoint for perceiving the body of others, as I mentioned above. At the suggestion of Gérard Jorland, my group came up with a paradigm[23] along the lines of one developed by the psychologist Janet Bavelas.[24] To get a sense for it, try the following experiment:

—Tell someone to stand facing you.
—Lift your arms at your sides until they are horizontal. Ask your experiment partner to do the same.
—Say to your partner, "Do what I do as quickly as possible."
—Rapidly place your arms at an angle, like the hands of a clock (right arm lower, left arm higher).
—You will observe that, most often, your partner will imitate you in a mirror fashion, that is, if your right arm is pointing toward the ground, he will point his left arm toward the ground.
—Now say to him, "No!" and keep your right arm lowered. You will note that, once he has gotten over his surprise, he will change sides. He will mentally put himself in your place.

His first response is to mirror your movements. But then he will switch to "rotational symmetry," which we believe to be a behavior typical of an empathic process.

To investigate these two strategies experimentally, we used a screen image of a tightrope walker created by the artists Michel Bret and Marie-Hélène Tramus, experts in digital images. The subject faced the virtual tightrope walker. The walker held a horizontal bar. We asked the subject simply to move forward or backward, and we observed whether he was imitating the model in mirror fashion or in rotation. Sometimes, we instructed the subject to imitate the tightrope walker as in the experiment described above. A recording of the subject's electroencephalographic activity showed that different brain networks are activated during the two strategies, which we believe to correspond to sympathetic and empathic behaviors.[25] We recently used this experiment with schizophrenic patients, in collaboration with the psychiatric service at the Poitiers hospital, and demonstrated that some of these patients have difficulty putting themselves in the place of the tightrope walker.[26] These findings may suggest difficulty in these patients' ability to relate empathically to others.

These strategies are most likely used differently depending on the psychological state of the subject. This work highlights the importance of vicarious processes in social relations.[27]

Suppressing Emotions to Help Others

But there is still more to say about empathy. It is hard to help a person who is suffering if you really have to feel that person's pain. If doctors were to suffer along with their patients, they could not help them. One has therefore to inhibit the emotion induced either by emotional resonance (sympathy) or by the change in perspective. Empathy is thus a very dynamic process whose mechanisms are sometimes antagonistic. Some of these mechanisms make it possible to change spatial perspective and to embody someone else, whereas other mechanisms allow us to adopt a fuller perspective at a remove from the situation, and to produce prosocial behaviors, that is, that lead us to help others. It is an ever-changing collaboration and competition between many brain networks. The future will probably show that several levels of this remarkable capacity have developed during evolution and that, as is the case with so many cognitive and emotional properties, the singular will be replaced by a plural, and we shall speak of the many aspects of empathic processes and their diverse pathologies.

8

Vicarious Emotions

The concept of vicariance has been used in many fields. This short chapter describes a few of these extended uses in order to illustrate its richness and potential extensions.

Vicarious Pain

"Vicarious pain" is a term used in psychiatric pathology to refer to the genuine impression of pain that a person feels when observing the physical suffering of another. Here, it is not the other person who is the vicar; it is the observer herself who takes the place of the other. This phenomenon represents yet another variant of the concept of vicariance. For example, patients placed in a functional magnetic resonance imaging machine are shown pictures of people cutting themselves or pinching their fingers in a door. Such experiments have revealed which brain network is involved in perceiving the pain of others. Observing the pain of others influences alpha-cortical electroencephalographic activity. Different regions are involved in perceiving one's own pain and that of others. Consequently, it is possible to separate the affective aspect of perceiving the pain of others from the purely sensory aspect. There are major differences between the sexes and among individuals vis-à-vis the capacity to feel the pain of others. These differences can also be strongly influenced by the social context (war, for example, requires that one minimize sensitivity to the pain one is inflicting on the enemy; love, or compassion, that one maximize it).[1]

Vicarious Traumatization

The notion of vicarious traumatization, also known as secondary traumatic stress disorder, designates a very curious pathology that affects people who are involved long-term in relieving the physical and psychological suffering of others, including nurses, social workers, rescue workers, and firefighters.[2] For example, humanitarian workers derive great satisfaction from their work, but it can also induce in them a series of symptoms that correspond to the notion of vicarious traumatization. The greater the degree of motivation and responsibility, the more serious the symptoms.

These symptoms include feelings of injustice, weakness, and various forms of anxiety that may present as hypervigilance, cardiovascular disturbances, disturbed sleep and nightmares, and troubled social relationships that may lead to isolation. Sometimes "compassion fatigue" appears in those who continue their efforts with the suffering. It is not just simple fatigue, but a more profound disturbance characterized by lack of motivation, distrust, and irritability. It is sometimes very difficult for these people to recover their equanimity, for example, in the case of war veterans. It is as if the traumatization reaches to the very core of one's personality.

This illustrates the use of the concept of vicariance in the specific sense of "standing in another's shoes." The affected people are so empathic that they cannot activate one of the basic mechanisms of empathy, which is the ability to inhibit feeling the others' sorrow in oneself. Recall that empathy refers both to the capacity to feel the emotion of others by putting oneself in their place and suppressing it in oneself so as to be able to help the other person. Empathy is obviously also involved in sharing joy. It enables one to adopt what the current scientific literature refers to as "prosocial" behavior. Finally, note that the ability to inhibit emotion felt by others also allows one to avoid being drawn into collective madness—crowd phenomena— triggered by the resonance that sympathy arouses. Empathy is thus a process of cooperation and competition that permits the genuine comprehension of others as well as the ability to remain independent. In this sense, empathy is a powerful source of vicariance.

Attachment

A good example of the capacity to consider another being as oneself, which is not specific to humans and is widespread among nonhuman primates, is attachment. One of the most compelling signs of attachment is grooming: "Grooming was the best single class of interactions for revealing the network of social attachments within the group [of rhesus monkeys]."[3] Three areas of the brain are important in these types of interactions: the orbitofrontal cortex, the anterior temporal cortex, and the amygdala.

Ostracism Hurts

Living in a group requires flexibility of roles and constantly changing relationships of sympathy and empathy; in other words, a supple social vicariance. When an individual does not adapt to the group, she is often excluded and ostracized. Now, recent research has shown that ostracism leads to the desire to be reincluded in the group, and thus to a greater tendency to imitation, but also to antisocial behavior that may turn aggressive. Ostracized people are attracted to extremist groups because they feel accepted there. The most spectacular finding in neuroscience is that the network activated by ostracism is the same as that activated by physical pain.[4] Exclusion hurts physically, and if I may hypothesize, this is not totally alien to the sort of distress that can lead to suicide. When a worker is subjected to bullying by management or the hierarchy, or subjected to cruel and frequent shifts in her professional identity, she experiences a particular kind of ostracism, social isolation, and loss of identity.

Altruism and Shame

Altruism is obviously a vicarious process. Social psychology tells us that altruists wish to be agents of others' well-being. Altruistic behavior is associated with activity in the cerebral areas of the brain that are involved in agency, the feeling that one is the author of one's own actions.

Shame derives from vicarious processes in two ways. If a member of a group behaves in a shameful manner, the group as a whole may feel implicated and consequently ashamed.[5] The group replaces the individual. Another way of conceptualizing contagious shame is personal. If a person is empathic, putting herself in the shoes of a person who is behaving shamefully or who feels ashamed may cause her to feel ashamed herself. The areas of the brain activated by shame are now known. They are the anterior cingulate cortex and the parahippocampal gyrus.[6]

Lying and Loyalty: Respecting Social Norms

The relationship with one's vicar ought to be one of mutual trust and loyalty, devoid of lies. The neural basis of lying and loyalty is a specific area of investigation within the cognitive neurosciences. New research models have provided access to the neural basis of the capacity to detect whether others are being honest with you or lying to you.[7]

Feelings of loyalty and lying have been studied using models developed in game theory and in experimental economics. For example, one model used a modified version of the famous "prisoner's di-lemma," which involves denouncing a partner to save one's own life. Other models concern the desire for revenge. In this way, pain is induced in the context of a game. Findings have shown differences between women and men, who change their empathic brain responses toward pain when they are playing with disloyal partners. If a traitor is subjected to pain, men show increased activity in the areas of the brain related to reward, and this activity is also associated with the desire for revenge.

Respect for moral rules and social norms is very topical and is being studied, for example, by Sylvie Berthoz, both in psychiatric patients and using brain imaging.[8] This work reveals complex networks that include the orbitofrontal regions. Modern methods have revealed differences in socio-affective behavior.[9] Research in social neuroscience on respect for moral rules also touches on delicate subjects.[10] An example is the following thought experiment: It is wartime. You are hiding in a cellar with several other people. Enemy soldiers are outside. Your baby starts to cry very loudly, and if you do not do anything, the soldiers will find you and kill you—you, your baby, and everybody else in the cellar. The only way to avoid that is to cover the baby's mouth. But if you do that, the baby will suffocate and die. Is it morally permissible to do that? The prefrontal cortex is involved in this terrible thought process by inhibiting nonmoral behavior and, as we saw in the work of Étienne Koechlin, arbitrating difficult decisions.

Part IV

Education

9

Vicarious Learning

[The minor rationalism of the twentieth century] was the fossil of a major rationalism (that of the seventeenth century), rich with a living ontology, that had already died out by the eighteenth century, and only a few external forms of which remained in the rationalism of 1900.

—MAURICE MERLEAU-PONTY

The triumph of across-the-board rationalism in the twentieth century gave rise to learning theories that ignored many of the qualities of the learner. The diversity and creative capacity of the living apprentice were forgotten, as happened in the industrial world. In services, the living agent was cast aside. Yves Clot has written widely on this topic.[1] One of the consequences of narrow rationalism was to standardize learning and to disregard the diversity of the learners.[2] As a result, this dimension now needs to be reintegrated. That could be done with the aid of vicariance, in the sense of Antoine de La Garanderie or Maurice Reuchlin, that is, the freedom to discover one's own solution to learn a piece of knowledge, a skill, or compensate for a deficit, or in the sense of von Uexküll, the ability to transform the world based on how one intends to use it.

The Legacy of Comparative Psychology

I have already noted that flexibility is a fundamental property of the human brain, not in the strict sense of suppleness, but rather in the ability to switch strategies. The psychologist Michel Huteau emphasizes a fundamental aspect of Reuchlin's theory: "Note that such a conceptualization is at odds with the rigid descriptions of inter-individual variability offered by classical comparative psychology, where the subject was considered to have stable characteristics. Here the inter-individual variability of behaviors is studied in conjunction with their flexibility. Note also that individual factors are taken into account in the same way as situational factors and in interaction with them. Vicarious processes can also be antagonistic insofar as they do not lead to absolutely identical responses."[3]

In other words, vicariance is a powerful tool of adaptation and learning for each individual throughout her personal experience. These ideas were picked up by Reuchlin's successors, such as psychologist Théophile Ohlmann, who also investigated the differences in behaviors within individuals (intra-individual). He writes: "In fact we believe that intra-individual variation is a genuinely transformative agent that permits a group to function differently depending on situational constraints and to switch from general laws to specific laws and vice versa. Combining these two categories enables a group of people to adapt to almost any significant change in the environment."[4] Note that Ohlmann emphasizes intra-individual differences, whereas the preceding citation about Reuchlin's theory refers to inter-individual differences. Both of these kinds of differences constitute the field of study of vicariance. If Antonin Dvořák's great cello concerto, which I adore, does not affect you the way it does me, it is because I can hear in it many different things with each listening, and also because we are different listeners and bring to the music our own cultural models, sensibilities, expertise, ignorance of musicology, and so forth.

In other words, for Ohlmann, vicariance produces effects not only at the level of the individual but also at the level of collective behavior, because distinct individual approaches to strategies lead to marked inter-individual variation. Variation among individuals creates the rich variations that exist among groups. This idea is very important because it challenges the norm. Suppressing the freedom of vicarious choice stifles and undermines the group; it crushes and standardizes creativity to the point that people begin to complain that innovation is dead!

I suggested earlier that these ideas are directly applicable to methods of rehabilitation and learning following brain damage. There are many methods, but some of them do not allow for this freedom. For example, robots used in rehabilitation often impose constraints that limit the brain's capacity to innovate. It is important to avoid machines that produce very stereotyped movements, in favor of re-habilitation by trial and error.[5] One finds the same problems in learning technologies as in teaching: there is a tendency to impose a single mode of learning, when the brain should be allowed the freedom to call vicariance into play. This is also the power of the sci-ence education movement called La Main à la Pâte that was launched in France by physics Nobel laureate Georges Charpak and is being adopted in Italy through a joint initiative of the French Academy of Sciences and the Accademia Nazionale dei Lincei, as well as else-where in the world.

The freedom accorded to a subject to find her own way in com-pensating for deficits requires acknowledging individual diversity. It is also valuable in education more generally. A Unesco report on cultural diversity in education mentions the notion of vicariance: "Taking into account cultural diversity means taking into account the diversity of populations, as well as the diversity of ways of teaching and learning. . . . It is in this spirit that we analyze the transition from differentiated instruction to differentiation in modes of in-struction. It is also in this spirit that we emphasize the basic distinction between an approach defined by categories and profiles to the detriment of the principle of variation through the notion of vicariance, for example. . . . Different styles of learning, but also differences in personal history, experiences outside of school, and individual knowledge—all these differences can no longer be ignored by schools."[6]

Learning by Vicarious Reinforcement

The concept of vicariance has been used to describe several forms of learning that involve, to various degrees, observation of others. One primary form is learning by vicarious reinforcement. This notion stems from the classic finding that learning can be improved by reinforcement, for example, by reward or punishment. The neurobiological and neurochemical mechanisms of this effect are now partially known. They involve not only purely electrophysiological mechanisms such as "long-term potentiation," but also the role of emotions and motivation through neuromodulation due to the influence of dopamine.[7]

The effect of reinforcement can also be induced by cognitive activity indirectly related to the concrete realization of the action that is the object of learning. This has been described in the context of so-called fictive learning methods.[8] Observing others profoundly influences learning, and it may be a fundamental principle that recurs throughout evolution. Much research has been devoted to studying the effect of observing others in primates, mice, and rats. The English-language literature uses the term "vicarious reinforcement learning" to describe the favorable effect of observing others, which obviously is related to the concept of imitation. It is beyond the scope of this book to summarize these findings; suffice it to say that learning by observing others is a basic component of our capacity to learn.

Two main strategies have been proposed with respect to learning by experience. One is based on the value of past experience, and the other is linked to the triggering of behavior by the context or the environment. The same holds true for vicarious learning by observing others. It may be either linked to goals or simply induced by stimulus/response–type processes.[9]

Learning through Vicarious Activity

Mastery learning through vicarious activity originated with the work of Reuchlin and Lev Vygotsky and consists of modeling oneself on others. A form of learning by imitation, it involves associating more advanced children with other children, who are thus stimulated and are provided with a model for what to do. This is another illustration of the basic difference, discussed in Chapter 7, between sympathy and empathy. My own reading of the literature on vicarious mastery learning tends to associate this type of learning with an empathic process such as that espoused by philosophers Gérard Jorland and Bérangère Thirioux, for, according to its proponents, this method allows the child to identify with her model. In spring 2000, the Association Canadienne d'Éducation de Langue Française (Canadian Association of French Language Education) devoted a special issue of its scientific journal to the subject, and the French academics invited to participate made known their different views. For example, one of the topics of debate was "learning style." According to Reuchlin, each individual has a repertoire of unequally "evocable" styles depending on context, preferences, and dispositions, as Pierre Bourdieu used the term with respect to his notion of *habitus*. However, I do not possess the pedagogical skills needed to evaluate the advantages of this method.

Bandura's Theory of Vicarious Learning

In most theories of learning, what is called "learning" can only be done by accomplishing an action and experiencing the consequences. The educational psychologist Albert Bandura concedes this fact but suggests that learning through direct experience most often occurs vicariously, that is, by observing the behavior of others and the consequences they incur.[10] This saves the observer from having to learn by trial and error. Bandura also stresses the importance of the learner's perception of her own ability to accomplish a task. It is not only a question of degree of motivation or of imitating a teacher, as is promoted in France, for example, but of the feeling of "being able to" and having the self-confidence to succeed. This typically Anglo-Saxon attitude is, in my opinion, very constructive because it favors initiative.[11]

Although I am not a specialist in education, it seems to me that each person has their own way of learning. The challenge of teaching and learning is not to find the one best method, but rather to discover the one best suited to each individual brain. You might go so far as to say that, actually, each person must find their own method. This is a simplex principle. In other words, it would appear, a priori, that it would be more complex to let each individual find their own path, whereas in fact it is much more efficient to do so. This is basically the method followed in English universities, where the students have only tutors and are responsible for planning their own route. The same approach is in use at the École normale supérieure.

This is the crux of the difficulty that trips up all teaching reforms that aim to establish standards rather than treating the process of learning like a vicarious process, in the true sense of the concept: recognizing that there are many possible ways, and that it is important to view another person as a vicar who is there not to replace oneself but to provide an example. This application of the concept of vicariance is consistent with the thinking of Reuchlin and La Garanderie, that is, that each individual has her own method. It is also an example of vicariance in the sense of Bandura, that is, that learning by observing others, and not solely through imitation, is a useful adjunct for certain minds.

Modern research on new methods of teaching based on the use of digital tools, without replacing the role of the teacher, opens fascinating perspectives, especially due to the presence in digital environments of avatars whose behavior the learner can observe. These tools promote vicariance by allowing each person to find a learning strategy that best suits them. Recently, in a small and very remote village in Ethiopia, illiterate children were given twenty or so digital tablets containing a few reading software tutorials. The children pounced on these tools and learned the alphabet by themselves, very quickly and individually, but with collective delight. It was obvious how every child embraced the tool and adapted it to their own skills and dispositions, since no method was imposed. Obviously, expanding these technologies to entire resource-poor countries poses a financial challenge.

Vygotsky and the Effect of Social Context

Like Bandura, the Russian psychologist Lev Vygotsky was also a defender of the importance of learning by interaction with others.[12] However, his theory is much more general, since it suggests that there is cultural and social anticipation of the ability to learn in children. According to his concept, which he called the "zone of proximal development," the collective and social example helps a child to anticipate and to some degree prepare for the effects of biological development. Here vicariance involves not only imitating another, but being able to create a context that makes it possible to anticipate biological development. Sociologist Nathalie Bulle writes: "Vygotsky's psychology relies on the idea that the rational faculties depend on the control and regulation of cognitive processes at a conscious level, thanks to the use of cognitive 'tools' that mediate thought. This is why it subordinates the cognitive development of the child to the child's progressive appropriation of cultural constructs, and especially organized constructs such as language and conceptual systems."[13]

Comparing the theories of Piaget and Vygotsky, Bulle maintains that, according to Piaget, the stages of development are linked to biological mechanisms that come "from inside," in other words, from the development of the brain, whereas for Vygotsky they are "drawn from without." To my mind, these two approaches are complementary. Vygotsky criticized Piaget's "linear" conception of development; he disputes the idea that there is a progressive transition from a solipsistic and egocentric perception to a decentered one. This debate is summarized in the work of the psychologists Anastasia Tryphon and Jacques Vonèche.[14] With respect to vicariance, what is key is that Vygotsky emphasizes the formidable contribution of culture and of social interactions in vicarious processes, however diverse they may be.

La Garanderie's Theory of Mental Movement

Reuchlin's theory of vicarious functioning assumes that each individual can resolve a problem in different ways, but also that different individuals have their own personal cognitive strategies. The theory also challenges the domination of language to leave open the process of learning based on individual talents. It is this aspect of learning that the psychologist Antoine de La Garanderie has addressed.[15] La Garanderie describes a typology of children who are, according to him, either more "visual" or more "auditory"—that is, better able to understand and solve problems if they are posed in a way that corresponds to the child's internal processing mechanisms. La Garanderie posits that the young brain uses "mental images" (hence the term "mental movement," which favors one or the other of these sensory modalities). This is why he objects to methods that force children to solve problems in a standard way, for example, solving a geometry problem simply by looking at it on paper. In this case, La Garanderie says to the child: "Put the problem in your head," thus stimulating the child to reformulate the problem according to her own preferred strategy.

On greeting a child, La Garanderie asks her how she imagines the path from her home to school, and thus distinguishes differences in each learner's methods. According to La Garanderie, there are three kinds of students: the editor, who edits very quickly but is rapidly limited in the ability to reflect; the seeker, who has no plan and often ends up confused; and the planner, who comes up with a beautiful structure but cannot always manage to flesh it out.

Like Reuchlin, La Garanderie uses the concept of evocability: the problem has to be "conjured up in the head." One of his examples involves how to comprehend the notion of a triangle. A "visual" student sees the triangle in her head and has no difficulty drawing it. An "auditory" student will be completely disoriented if she is told that a triangle is "the union of all the points of a plane bounded by three secants," which requires constructing a mental image of the shape. This student prefers a sequential presentation of the problem. If you ask her to draw two straight lines that intersect, and then to draw another line that intersects the two previous ones, and finally to color the interior of the figure, she will be able to remember the sequence and to perceive a triangle.

I mentioned a personal anecdote of this form of vicariance in Chapter 4. When I was in the preparatory class for the "grandes écoles" at the Lycée Saint-Louis in Paris, we had to take a test in descriptive geometry. It was very hard for me to solve a complex problem in geometry that required calculating the intersection of a hyperboloid and a paraboloid. My classmates succeeded in solving the problem using algebraic methods, but I could not. However, I was able to find the solution by resorting to imaginary points, called cyclic points. How is that for a fine example of vicarious mental strategies? My brain was more comfortable in virtual geometric space than in the written sequence of algebraic equations.

Another example from La Garanderie is squaring a binomial. A student whom La Garanderie characterized as "auditory" learned the following formula by heart:

$$(a+b)^2 = a^2 + 2ab + b^2$$

For a "visual" student, however, one could suggest a drawing made using these instructions: draw a square with each side having the length a + b, then draw small unit squares within the large square. The result will immediately be evident.

Teachers themselves have preferences. La Garanderie collected academic profiles of both teachers and students. We often forget that every teacher is different, cognitively speaking. Yet it is essential to adjust the preferences of teachers' cognitive strategies to the capacities and the preferences of the children.

In terms of teaching, we have also seen the domination of one-size-fits-all methods, as if mathematics, physics, and geography could be learned by everybody in the same way. In launching the initiative La Main à la Pâte, Georges Charpak not only reintroduced action as a basis for understanding physical concepts but also emphasized the diversity in individual methods for internalizing concepts. Similarly, Bertrand Schwartz, an engineer by training, taught the idea of coupling forces to untrained individuals by means of direct experience. Jean-François Billeter's book *Leçons sur Tchoung Tseu* [Lessons from Chuang Tzu] contains wonderful ruminations on the role of action, the body, and movement in Chinese thought. For example, Billeter cites this quotation from Chuang Tzu: "Oh! If only I knew a man capable of forgetting language, so I would have someone to talk to!"[16]

Cognitive Modification of Learning

I described above some examples of solving the same problem in many different ways, which corresponds to vicarious functioning.

But what are the factors that might influence the choice of this or that process? Drawing a distinction between visual and auditory people (Jacques Paillard further distinguishes "proprioceptive" subjects) emphasizes innate factors or learning at a very early age. But cognitive factors such as context or "cognitive priming" cues can modify the adoption of this or that strategy. Below are a few examples.

Vicarious Learning in Context: Dual Adaptation

Suppose that you are traveling in England and have been driving on the left side of the road for a week. This situation requires a profound reorganization of your way of driving. The reflexes acquired to drive on the right are called into question, and your brain must construct new visuomotor coordination patterns. Suppose, now, that you return to America after this trip. Your brain must rediscover its former coordination patterns; often errors occur that could lead to a serious accident. The new coordination acquired in England persists for a long time and emerges when you least expect it. For example, when you go to turn at an intersection, instead of staying in the right lane, as you should, you are aghast to find yourself taking a new direction in the left lane as you would have done in England! This shows that the brain has acquired and memorized two different modes of driving. This is known as *dual adaptation*. It is a form of vicariance, according to the definition of vicarious functioning, since two modes of accomplishing the same task coexist.

There are many occurrences of this type of vicariance. For example, after a long time at sea, sailors feel as though the ground is moving under their feet. This is due to the fact that the brain constructs a dynamic pattern of oscillating neuronal activity when one is at sea that is in counterphase with the movement of the water and thus gives the sailor the impression of being immobile. This dynamic pattern is profoundly rooted in the functioning of the brain and persists even after the return to terra firma, giving the impression that the ground is undulating!

Sensory conflicts are another example of vicarious adaptation. For example, wearing reading glasses induces an adaptation between visual and vestibular information, watching a waterfall for a few minutes causes the well-known illusion of a moving landscape when you look away from the water, and so forth. Astronauts who in the future will spend long periods in microgravity may lose their adaptation to gravity. One idea being seriously considered is to equip the vessels that will carry them to Mars with centrifuges so that they will be able to maintain their adaptation to gravity at the same time that they are adjusting to microgravity or to Martian gravity. The centrifuge will generate a force that will simulate gravity and enable the astronauts to preserve their adaptation to terrestrial sensory motor conditions.

Dual adaptation depends on the context. For example, subjects can store two sensory motor configurations and can alternate between these two modes based on contextual cues around which each configuration is formed.[17] My colleagues and I have explored this "vicarious adaptation" by means of a model of a conflict between vision and kinesthesia, and more precisely by modifying visual information as the body rotates during virtual navigation.[18] We project into a virtual reality helmet a virtual corridor furnished with several twists and turns through which the subject must navigate, as if she were moving through the subway. We create a device that modifies the direction of the path in the virtual world based on the subject's movements in her chair. Normally, for example, to turn 90 degrees in the virtual world, the subject would have to turn 90 degrees in her chair. But we created a conflict by changing the relation between the subject's movements and the virtual rotation. Thus, to turn 90 degrees in the virtual corridor, the subject must turn 150 degrees in her chair.

In an initial experiment, we asked how subjects would adapt to the conflict. Would they favor visual information or kinesthetic information? Neurophysiologist Jacques Paillard, and later his student Michel Lacour, showed that some people are "visual" or "proprioceptive." The results of this experiment show that, in fact, people fall into two groups that use different strategies corresponding to two models of sensory dominance. Several strategies are thus possible, involving various combinations of neuronal networks. These strategies are present in potential (Reuchlin's "evocability"), with their activation depending on context, training, and so forth. These findings are consistent with La Garanderie's hypothesis regarding the dominance of the use of certain senses to solve cognitive problems.

In a second experiment, we trained subjects to navigate through the corridor under two conditions, C1 and C2, with different relationships between vision and the rotation of the body, thus creating a model of dual adaptation. For example, the relation of the angle of rotation in the visual corridor and in the chair, respectively, was 90 and 120 degrees in C1 and 90 and 150 degrees in C2. The hypothesis was that the brain is able to code the two modes of cooperation between vision and proprioception, but that to tell them apart, it uses contextual cues, as you do when you are driving in England or America.

Here, the general idea is that the brain links different sensory motor modes on the basis of context. The conclusion of this experiment was that it is possible for the brain to adapt and to encode two modes of adaptation to a sensory motor conflict between visual and kinesthetic information about the body's rotation during a task of navigation in virtual reality combined with body movements. This dual adaptation is fast, but it depends on the type of conflict imposed. The use of contextual cues to induce one or the other adaptation cannot be explained by simple models. To explain the variations observed among subjects, it appears to be necessary to add supplementary processes, such as factors relating to the stabilization of the internal model that is continually in use.

The Effect of Cognitive Priming on Kinesthetic Learning

Another mechanism for varying sensory motor strategies is *cognitive priming*. Imagine that your eyes are closed and that you would like to walk in a path that describes a square. To do that, you "put in your head" the image of a square, and you let the image guide you as you walk. Now suppose that, instead of a square, I ask you to walk a path describing a diamond shape. You adjust the image and reprogram your movement to effect a locomotor path according to this shape. We have experimentally established the influence of this cognitive priming in the following way.[19]

Blindfolded subjects enter a room containing a hexagonal path marked by a ribbon tied to supports, like those used to control crowds in museums and airports. Subjects learn the shape of the path solely by walking along the ribbon; that is, they acquire a kinesthetic experience of the shape without any visual cues. After this learning, the ribbon is removed, and subjects are asked to reproduce the hexagonal pattern while walking. Subjects are generally successful. Then we have the subjects leave the room and we show them, on paper, a drawing of a different path—a diamond, say—and tell them that that they will be doing it in the dark (this mental operation is called cognitive priming). We then have subjects walk the hexagon in the dark, whereas they believe they are following a diamond. Finally, we ask subjects to reproduce the shape by walking in the empty room. The result is that they no longer walk the hexagon, but instead follow a path influenced by the shape seen on the paper.

The brain has a remarkable capacity to imagine an abstract shape and to transform it into motor coordination, aided by several effectors. You can draw a letter with your finger, your arm, or your tongue, or while walking. This is what is called the principle of motor equivalence. It is a simplex principle that derives from the fact that the brain uses very general geometries.[20] This enables it to formulate plans for movements in a very general way. These plans can then be implemented by different motor apparatuses. This is a very subtle form of vicariance, that is, the capacity to produce a motor form by different motor apparatuses, whatever their complexity, by virtue of a very general formulation.

Another example of this use of general variables for vicarious functioning is eye movements, which I have described in depth in other books.[21] The eye movements you execute to read this book are called ocular saccades: they are rapid changes of eye position that move your gaze from one point to another in the environment. Now, suppose that you wished to move your gaze from one spot in the room where you are to another. This displacement can be made with the eye, the head, or your entire body. The decision to look stems from mechanisms that have been the subject of much study.[22]

Although its goal is to direct changes in position, the brain uses speed as a control variable, or it even uses composite variables that blend speed, acceleration, and position, which makes it possible to orient gaze with the eye, the head, or the body, and thus to compensate for an eventual deficit of one of these motor systems. In this case, vicariance is not duplication (as with the vicar or the double); the multiplicity of possible solutions (the different effectors) is linked to the great generality of mechanisms with which evolution has endowed us, to enable vicarious functioning. These mechanisms are still largely mysterious.

Why are these experiments relevant for my theory of vicariance? Because they show that cognitive information profoundly influences motor behavior. Vicariance can be induced in a top-down way by cognitive factors. This suggests, moreover, that deficits that appear to be motor-related may be due to problems of spatial processing or memory. Consequently, compensating for these deficits must also be targeted, as I showed earlier with respect to mental imagery, toward these cognitive factors, which are the source of vicariance; whereas today, rehabilitation still emphasizes only motor aspects.

Epilogue

All the philosophical and political thought of
modern times tends to place the human mind
on a plane above reality, which creates an abyss
between man and the world.

—E. LEVINAS

It is time to wrap up this book. Do not read it as a treatise or a manifesto, for it is neither. Rather, its purpose is to open a debate intended to break the bottleneck caused by overly conformist ways of thinking. I would like to take advantage of the freedom afforded me by this book to discuss a few additional aspects of vicariance. Think of this epilogue as a bouquet of wildflowers like the ones you might gather while out for a stroll. I dare hope it will contain a mix of colors and shapes that will surprise and inspire you! Today we need to rethink so many dogmas, and to do that, we need to see clearly. This epilogue also has some aspects of a fireworks display. It aims to be a summary with éclat. The beauty of the final bouquet in a fireworks display is that it unfurls, thus indicating the beginning of something rather than the end. Indeed, I hope that this collection of variations on the theme of vicariance will add up to more than the sum of its parts.

The Many Facets of Vicariance

Now that we have analyzed the concept of vicariance, is it possible to discern the constants that assure its applicability in communities as different as those of paleontologists, educators, psychologists, and physicians? In fact, the essential message of this book is that vicariance is a fundamental property of life. It is more than a mere representation, or an act of delegating, or a proxy. Through a fruitful manipulation of its meaning, "vicariance" also denotes a substitution that promotes creation by means of diversity (which in itself is trivial) but above all by means of the productive tension between the universal and the particular.

Indeed, vicariance is much more than representation or delegation, which are only mandates. A representative, like an ambassador, has a task to do but has no freedom of action. Vicariance, on the other hand, always implies an act, an intention, a shared goal. It stands for the creation of new solutions. Moreover, the concept of vicariance always deals with the relationship between two individuals, two processes, or two mechanisms with the same goal.

Invention is generally an idea for replacing one technology with another, but it is often vicarious, that is, it brings added value. In replacing wheeled vehicles, tracked vehicles enabled movement over more hazardous terrain; in replacing carbon, petroleum gave rise to refined derivatives that are the basis of the chemical industry; and so forth.

Throughout this book I have shown the various facets of vicariance. Beginning with the substitution of the vicar for the priest, we have seen the concept evolve and its meaning change. Obviously, I might be accused of pushing the limits of the concept, but I specified at the beginning that I did not wish to constrain it. Indeed, my hypothesis is that behind this multiplicity lies one or more fundamental constants of life, relating in particular to the human brain and to social life. The omnipotence of living cells, especially certain stem cells that can substitute for each other, is astonishing. In this case, vicariance is not about replacement but about ambivalence. And the redundant multiplicity of systems that code for the same information could as well be called informational vicariance. Vicariance thus simultaneously implies specialization, modularity, and genericity. The most recent use of the term is in association with the creation of digital avatars, which offer the possibility of duplicating oneself and assigning tasks to the avatar. This innovation allows one to conceive of and produce avatars of oneself that have different properties. To Ricoeur's *mêmeté* (sameness) and *ipséité* (selfhood), one must add a third category, *avatar-ité*. But, at core, a key idea arises from the history of the concept: vicariance is a source of creativity and innovation. In the following sections, we will test this idea by further generalizing the notion of vicariance to other areas of human thought.

The Odd Couple: Universal and Particular

The concept of vicariance raises the question of the relationship between the one and the many. The uniqueness of life is that it straddles two extremes: the universal and the particular, the species and the individual, the rule and its exception. We are determined by our genes according to universal rules whose validity for all living species we bear witness to every day. Genetic variation leads to variations among individuals as well as to many pathologies. Subsequently, epigenesis profoundly modifies the functioning of genes at the time of their expression, based on the environment. A part of the history of humans results from this tension at all levels, from molecules to social interactions. All of family, social, and political life is a subtle and often violent play of private and collective interests.

We have recently left a century dominated by the impulse to standardize and the search for universal truths. Today we are seeing the vindication of the individual in the face of the immense power of social institutions. René Cassin, a French jurist who fought all his life for the right of any individual to question the state before the international courts of justice, understood this very well.

If standards are what allow us to escape the despair induced by the infinite variety of possible behaviors, it is departure from the norm—calling it into question, and frequently—that allows us to create new connections between the world and ourselves. "It is forbidden to forbid" was the rallying cry of May 1968. Today it is fashionable to disparage it or to consider it *dépassé*. Yet, to my mind, it was a harbinger of a historic change. The individual was demanding the return of autonomy in societies where conformity was based on obsolete criteria. This rallying cry was not an anarchical declaration but rather a desire to lift the lid on the decrepit rules of a stagnant society. Today we say that young people no longer respect social norms; what that suggests is that we are in the midst of a reformation of those norms.

For decades, mathematics has been taught in France as if the formalization introduced by the Bourbaki group (whose influence should not be underestimated) offers the only possible access to mathematic concepts. Yet, as a colleague told me, that is to deprive oneself of the richness of two methods for solving the same problem.[1] Science, too, is divided between two tendencies: showing the universal character of the properties and mechanisms of living things, and describing their diversity. Linguistics was, and still is, a battlefield where these two approaches face off. I do not deny the fundamental importance of Noam Chomsky's mission to discover a universal grammar, but it is contrary to the study and appreciation sparked by the diversity of language, so dear to the eminent linguist Claude Hagège. On a different note, Jean-Paul Sartre wrote, in *Search for a Method*: "The Marxist therefore is impelled to take as an appearance the real content of behavior or of a thought; when he dissolves the particular in the Universal, he has the satisfaction of believing that he is reducing appearance to truth."[2] And later in the same book: "We reproach contemporary Marxism for throwing over to the side of chance all the concrete determinations of human life and for not preserving anything of historical totalization except its abstract skeleton of universality. The result is that it has entirely lost the meaning of what it is to be a man."[3]

Serendipity: An Example of Vicarious Creativity

A particularly creative form of vicariance is serendipity. Pek van Andel and Danièle Bourcier define it well: "The art of finding things or the capacity to discover, invent, or create what has not been sought in science, technology, politics, or everyday life, thanks to a surprising observation."[4] Their book on serendipity provides many examples of discoveries made by chance, as a result of letting the mind wander or of the simple convergence of circumstances. It is banal to say that chance favors the prepared mind. But, actually, it almost always derives from formulating a strong hypothesis that is proved wrong by experiment, so that you have to come up with another hypothesis. This substitution is a creative form of vicariance. It assumes that the discoverer has the capacities described in this book: the ability to change perspective, seek inconsistencies in conventional rules, and challenge rules based on a single exception. The engine of discovery thus resides at the level of emotion: it is surprise, a function of the emotional brain, that alerts the cognitive brain. Emotion is an agent of vicariance.

You will forgive me if I play the philosopher for just a bit. When paleontologists use the term "vicariance" to describe the diversity resulting from climatic changes or continental drift, it seems to me that the designation "vicariance" reveals their specifically human capacity to analyze this diversity as the result of natural processes, and not simply the fact that these processes have led to diversity.

One must be capable of changing viewpoint, of becoming allocentric, of having a global external perspective—gazing down on continents—to identify the property of vicariance.

Vicariance is about breaking free of reality and its constraints in order to enter imaginary worlds that only the human brain can conceive (even though dolphins do dream, as neurobiologist Michel Jouvet maintains, as evidenced by the erections they have in their sleep!). Finally, vicariance allows us to accept others as different from ourselves and not only, as Ricoeur and Damasio have written, "the other as ourself." Vicariances (plural—since ultimately there is no singular concept that does not conceal a multiplicity of forms) are the foundation of multiple interpretations and, consequently, of tolerance.

Vicarious Actor or Spectator?

In the relationship between an actor and a spectator in the theater, catharsis, which, according to Aristotle helps the spectator to control his passions or even to suppress them via a symbolic transfer, is a very subtle form of vicariance. It is "vicariance in action." It is more than substitution or imitation. Through a detour by way of make-believe involving body and language, and through a powerful intersubjectivity between actor and spectator, it is a vicarious simplex process. It has several characteristics, in particular the fact that the exchange may take numerous forms depending on the people and the context. One could also speculate about the idea of transference in psychoanalysis, but that would be to go farther than I wish to.

Vicariance and Metaphor

Could metaphor be a form of vicariance? Actually, yes, it could, because a metaphor is a substitute for the real object, action, or person that it represents.[5] Today, the study of metaphor extends beyond the context of pure linguistics, especially in the light of the work of the cognitive scientist George Lakoff and his group.[6] A series of cognitive and neurological experiments have shown that metaphor is the product of several brain mechanisms linked to action and not only to language.[7] The ability to produce metaphors is disturbed in people with schizophrenia and autism.[8] People with aphasias of the right hemisphere are still adept in grammar and phonology but cannot invent metaphors. The linguist Jean-Luc Nespoulous showed that the absence of metaphor is detrimental to the comprehension of complex statements.[9] Metaphor relies on global capacities of the right brain that have largely been documented.[10] Also worth mentioning is the work of the neuropsychologist Peter Brugger on magical thinking, which is linked to the brain's ability to make loose associations. This property is essential for creating metaphors, in contrast with the left brain, which tends to associate words with closely related meanings. This corresponds well with the involvement of the left brain in the sequencing of action, in egocentric sequential path memory, and in the organization of language.

According to a recent review of data from brain imaging, the right cortex is activated in connection with processing a metaphor only when the meaning of the metaphor is new, or in the case of very special uses. The active areas of the brain are the right medial and inferior frontal gyrus, the insula, and the middle and superior temporal gyrus. It appears that the role of this part of the brain is to activate very large semantic fields and to integrate concepts that have only weak semantic links, which is very close to the role mentioned above of the right cortex in loose associations and magical thinking. The substantial amount of research done on this subject indicates that in reality, metaphor requires the cooperation of both hemispheres. This is not surprising considering all the varieties of metaphor: implied, extended, direct, lexical, dead, heuristic, humorous, Homeric, and so forth. There is no point in explaining each of them here; I will only cite a few articles from the abundant literature.

These different kinds of metaphors use a great variety of linguistic tools (nouns, verbs, adjectives) and evoke movement, or play on the senses and the memory. Thus it is natural that different parts of the brain are involved in each type of metaphor, although the right brain plays a key role. It is a common error to attribute to a specific area of the brain the responsibility for processing or initiating an idea while neglecting the variety hidden behind this apparent unity. Fairly recent work suggests the participation of areas linked to action in the production of metaphors,[11] and differences depending on the use of verbs or nouns, or reference to actions or objects.[12]

For the cognitive scientist Dan Sperber, metaphors do not transgress the norms of current language, but rather are "creative and evocative exploitations" because in one way or another, every statement resembles some thought on the part of the speaker: "On the whole, the closer one gets to the metaphor end of the literal/loose/metaphorical continuum, the greater the freedom of interpretation left to hearers or readers, and the more likely it is that relevance will be achieved. . . . So when you compare metaphors to other uses of words, you find a bit more of this and a bit less of that, but nothing deserving of a special theory."[13] Thus, metaphor assumes the acquisition of mental capacities for classification and, especially, of generalization. For me, metaphor is always rooted in action, even though it does require the capacity, acquired late in infancy, of abstracting from contingency.

Indeed, in children, metaphors are not understood or produced until after the age of four.[14] At this age, the child still understands words in their strictest sense. If you say to a child, "Daddy flew away on an airplane," the child will think that wings propelled his father. Around six years, the semantic components are differentiated; differentiation allows practical acquisition of analogies and images by creating semantic networks and wordplay. Finally, the mastery of metaphor occurs around eleven to twelve years of age, along with the capacity to change point of view. Metaphor is thus a detour, another—vicarious—route for expanding the senses. It creates novelties that are broader than the strict logical connections of ordinary language.

Catachresis

The notion of catachresis originally designated the misuse of a word to describe a situation or an object in a way other than the standard sense of the word: for example, the "arm" of a chair or the "head" of a nail. But it has also been used to indicate a process of substitution of meaning in work situations, when an agent uses one tool in the place of another or characterizes a situation using notions that have nothing to do with the task at hand. The occupational psychologist Yves Clot cites the example of an agent using one tool instead of another (e.g., using a screwdriver to open an electrical cabinet that normally would be opened with a key). One form of catachresis consists of designating a situation by modifying the meaning through use of different concepts. For example, when a hospital orderly is trying to relax a patient he is accompanying to the operating room, he may refer to switching gurneys as changing taxis.[15] Yves Clot distinguishes three sorts of catechresis in work situations: "The first are 'centrifugal,' that is, to a first approximation, turned toward action on external objects. Their primary function is to enrich the repertoire of tools to respond to circumstances of actual work. The second, called 'centripetal,' are destined, to a first approximation, to act on themselves to maintain a level sufficient for cognitive and subjective mobilization or to serve to call up new objects of thought. The primary goal of the third type is to enrich or to select the verbal lexicon or expression."[16]

What links the notion of catachresis to vicariance is the fact that the agent transforms a prescribed use of a tool or artifact (the screwdriver becomes a key) or modifies a prescribed process and tries another strategy to perform the same task (a truck driver gets around a speed limiting device by simulating a situation where the permitted speed is higher; a train conductor devises strategies to stay awake).[17] In Clot's words: "The artifact is not in itself an instrument or part of an instrument (even if it was originally conceived for that); it is established as an instrument by the subject, who endows it with the status of a means of attaining the goals of his action. He thus induces more or less major reorganization as a result of this action. But the same artifact may become a very different instrument depending on the subjects and, for the same subject, depending on the situation and the moment.[18] An artifact goes from its prescribed use to its effective use through the mediation of 'instrumental genesis.'" Catachresis thus provides evidence of creativity linked to rejection of the norm and the search by each agent for new solutions based on the context, his goal, his personality, and so forth. The subject re-creates his activity.

Vicariance and Society

Money

Money is a very subtle form of vicariance because it replaces something real. The invention of money can be described as reflecting a particular capacity of the human brain to find substitutes to accomplish a task, substitutes that respond to the principle of simplexity. Inventing money and getting it accepted collectively is not simple, but it is not complex, either; it is a great solution that I have referred to as simplex. It is a detour, in the sense in which I understand certain forms of simplexity. Like every detour, it can itself be detoured. The recent abuses that turned the financial industry into a world apart, having created a space for negotiation completely independent of reality, people, and goods, are an excellent example of the way in which the most remarkable properties of the human brain can lead to the most sordid excesses and to dramatic destruction of social life. Through corruption, money also ends up replacing humans. The concept of vicariance thus contains within itself a possible perversion. In agreeing to be replaced, the priest can be betrayed by his substitute. The student becomes the vicar of his master by usurping his knowledge. A Japanese proverb says that every student must rob his teacher. But he must also modify his thinking, and he can betray it.

Different Societies with a Similar Milieu

Writing about Amerindian civilizations, the anthropologist Philippe Descola observed: "The humid tropical forest . . . in the Amazon shelters relatively egalitarian societies that practice slash-and-burn cultivation. The societies are barely integrated at the institutional level and ignore social hierarchies. . . . In Mexico, in a somewhat analogous setting, the Maya figured out how to build city-states, with specialized castes. . . . Thus, using similar techniques, the same environment can give rise to very different political forms."[19]

This form of vicariance also applies to the evolution of societies when the environment changes. In the future this will become dramatically relevant with global warming, as has been the case during major climatic changes in times past. I call this anticipatory vicariance: humans are (or they should be) anticipating a reduction in the availability of fossil fuels and so, having already replaced coal with oil, we should use the threat of future warming to begin replacing oil with solar, hydroelectric, or wind power.

In short, it is this similarity in diversity that is called biogeographical vicariance or ecological vicariance and, for humans, social vicariance. In other words, the tension between universal properties and the diversity linked to continental drift and to many climatic and environmental factors has had a powerful creative effect. The concept of vicariance implies this creative power. While one might object that the choice of the word "vicariance" was made by pale-ontologists for want of any other, I would answer, following an Aristotelian path, that what matters is the meaning and action that we have the potential to bring to it. To be unwilling to attach to a concept more than one meaning is precisely to miss what this book is about: the capacity to create.

The Scapegoat: A Vicarious Substitute?

Let's consider the case of the scapegoat, a subject about which the scholar René Girard has written extensively. Girard's theory is that the scapegoat keeps one group of people from massacring a second group through the sacrificial substitution of itself for this second group, thus channeling the violence that Girard sees as a key element of relations between human communities. This violence is the result of a "mimetic desire" to have what someone else has. Girard pushed his theory to the extreme in stating that Jesus serves as a universal scapegoat through his exemplary sacrifice, thus making it possible to lessen the dramatic consequences of the mimetic desire by ap-peasing relations between people. Jesus the vicar was sacrificed by the whole world. The scapegoat is a vicarious agent that represents not just one person, but a group of people. Many anthropologists criticize this theory for want of empirical data to support it; but this is out of my realm of expertise.

Democracy and Vicariance

Democracy makes broad use of vicarious identity. If the vicar occa-sionally replaces the priest, the representative is a permanent replace-ment for the citizen. Representation is "vicarious politics." The notion of representation is, in fact, a late invention in human societies. It appeared when people approved the need to install intermediaries—representatives—between themselves and the absolute power of kings and chiefs, who throughout history have always ended up abusing their power. Moreover, as evidenced by the House of Atreus, the Borgias, the kings of England and France, or the occupants of the Kremlin in the heyday of communism, the mighty are more preoccupied with killing each other to stay in power, and allying themselves with prelates like the Italian popes and the cardinals of the Renaissance, who were themselves preoccupied with lucre, than they are with fulfilling their role as shepherds of the people and the poor. This process continues even today. The Arab Spring was a prime example of a popular rebellion against presidents such as Hosni Mubarak of Egypt, Muammar al-Qaddafi of Libya, and Zine El-Abidine Ben Ali of Tunisia, whose misuse of their charisma simultaneously destroyed the freedom of their people and the wealth of their countries.

Much remains to be done. Take another example. Recently, an official in Tunisia was interviewed on television by a journalist. The official justified the religious orientation of the state not only because the people voted for it but also on the following grounds: "France is a Catholic country; Tunisia can be an Islamic country." However, he seemed to be unfamiliar with French history and the fact that the revolutionaries of 1789 ended up renouncing the idea of a religious state, and that it took almost two hundred years and many ups and downs until a genuinely secular republic was established. He also seemed to neglect the fact that this secular republic permits a diversity of beliefs, and of the mental worlds and customs they engender, to cohabit in mutual tolerance. What is at work here is vicariance, in the sense of the freedom to allow the human brain to create individual worlds without imposing them on others. This is the distinctive feature of "interpretive pluralism."[20] At the same time, the secular republic establishes rules for sharing in social life that avoid the religious proselytizing and ethical laxness that are typical of sects. Liberty doesn't mean libertarian.

Substituting one person for another is useful in avoiding conflicts that, in the past, were settled by the sword or by war or armed skirmishes. It was a time when battles between chiefs could replace mutual annihilation. Sacrifice, as Girard describes it, could give a scapegoat the function of replacing combat between groups, by acting as a substitute. In this same way, a lawyer substitutes for the opposing parties in a dispute. It is a very original form of social vicariance, for, contrary to the scapegoat, the lawyer is very well paid for the role of remaining neutral. In some cases, communications between lawyers may be completely confidential, that is, they are not shared with the opposing parties. The appeal of a vicarious process in social transactions becomes obvious: the use of a replacement makes it possible not only to defuse emotions but also to change the nature of the relations between belligerents. Here, vicariance does not mean imitation, but a process of negotiation that leads to creative solutions.

Utopia: The Creation of Other Possible Worlds

Utopia seems to me to be one of the most general forms of vicarious creativity. Delving deeply into it is beyond the scope of this book.[21] I only wish to suggest here the relevance of the idea of vicariance in the brain's creation of worlds. Consider the novel by Ludvig Holberg that describes the adventure of Niels Klim.[22] Propelled by an explorer's curiosity, this young man decides to go search out a rift in a mountain. At the end of a vertiginous fall, he discovers another world situated in the depths of the earth. This world is populated not with humans but with trees that are a vicarious representation of humans. They speak, act, and move very slowly, and this alteration in timescale triggers profound social changes. The trees have built a society that is at once similar and different from ours. In fact, as often happens in utopias, its rules and laws turn the society into a political project of democracy that, while being inspired by the human world, alters its basic elements in a more democratic direction. This is an example of social vicariance benefiting from the creative freedom of the human brain to imagine other worlds.

Another example of social vicariance of this type involving politics is the book *Magellania,* originally written by Jules Verne and his son Michel,[23] which was so magnificently staged by Ariane Mnouchkine with the Théâtre du Soleil. In this dual history, a man and his wife dream of making a movie about a great human adventure and turn the attic of their house into a film studio. The studio is used as a tool of vicarious mediation, going from one world to another through the movie. The movie recounts the sea voyage, around 1881, of a capitalist in search of gold that he had heard was to be found in the distant and mysterious countries of South America. A storm and grounding in the Strait of Magellan (whence the title of the book) isolates the passengers and crew on the frigid and remote Isle of Hoste. They are saved by a local, Kaw-Djer, who (to my mind) symbolizes the link between humankind and nature. The group's situation leads some to propose an egalitarian community combining socialist and communist utopias that contrast with the appetite of the capitalist bourgeois gourmand. Several utopias clash. Here, too, in a dream, the brain as creator of worlds puts utopia to the test of reality. Vicariance not only creates a possible world but tests this world with the implacable breath of reality.

The Universal Declaration of Human Rights gave the linguist Emmanuelle Danblon the opportunity to discuss "the role of the imagination in the service of a specific rhetorical function: the expression of utopia."[24] In this context, she wrote, she had "proposed a strictly 'utopist' reading of the first article of the Declaration ('Men are born and remain free and equal in rights')." Yet, "such a statement could not logically be interpreted as responding to an epideictic need to *imagine* a world that did not exist or a lost paradise. This would obviously be counterproductive in a culture of human rights where liberty, equality, and dignity are values. Here, the role of the formulation is rather to suggest a *possible, desirable, and plausible* world. . . . In this case, utopia is a spur for action; it is different from decision making. If the utopia is never completely achieved, the horizon that it opens is as clearly articulated as a *beacon of hope.*"[25] I have emphasized this last phrase because vicariance is a bearer of hope. The capacity to switch perspectives, to find new solutions, to imagine and create possible worlds through adroitness and action is synonymous with hope. Ulysses's cleverness in finding his wife and his country, the patient who recovers lost function, and the math-challenged child who nonetheless finds a way to master mathematics based on his own skills—all of these exemplify vicariance as hope. It is despised by dictators, fanatics, consultants obsessed with production and easy money; it is the opposite of Taylorism. *It is a source of pride.*

Vicariance and Resilience

Resilience, or the recovery of function following different kinds of trauma, is a remarkable faculty of the human brain. The pediatrician Alexandre Minkowski and his interdisciplinary team demonstrated it in children who were victims of war or of terrorist acts, in Rwanda, for example, or of natural disasters, as well as in thousands of children locked away and isolated in institutions with no emotional contact under the Ceaucescus in Romania. Many authors have published personal accounts and extensive studies on this subject. Resilience is nevertheless, in my opinion, only one of the manifestations of vicariance, a general property of life whose possibilities and forms I have sketched out in this book. This issue merits discussion, of course, and I propose it very cautiously given the complexity of the two concepts. But we are talking about a work in progress, and readers who have stayed with me thus far will no doubt already have plenty of questions!

Doctor Jekyll and Mister Hyde: The Two Sides of Vicariance

The possibility of creating worlds is a marvelous faculty of the human brain, and multiple interpretations go hand in hand with it. I showed above that according to von Uexküll, each species constructs its own world. And each of us can construct many of them. Humans can also live in one world while acknowledging that others live in different ones. This is the basis of tolerance, of "cultural and social empathy," and ultimately of democracy.

Yet there is another side to this capacity to create different worlds. The risk is that each person may become fixed in ideas that are incompatible with those of others. Jean-François Billeter, a specialist in Chinese literature and philosophy, touches on this question. "Many worlds implies many conflicts," he writes.[26] "Objectification [is] an involuntary operation by which, through an imaginary synthesis designated by a word, we make something that we assume to exist objectively as we imagine it."[27] Note that "objectification" has a meaning opposite to "objective," which generally refers to the real world. Here, the word denotes a process of constructing a world of objects perceived by each individual. Billeter uses this definition to discuss the difference between "worlds" and "reality": "I purposely make a distinction between *world* and *reality*. By *world* I mean all the things among which we live, created by objectification and language within reality, and by *reality* everything that exists within and outside of us, independent and beyond the forms created by objectification and language. This distinction shows that within a single harmonious and homogeneous reality, we can live in *different* worlds. . . . Diverse societies can live in different worlds, even if they are incompatible."[28]

I have previously described the related paradox of projective perception. This is a simplex principle because it allows the brain to impose interpretive rules on perception, thus making it easier to analyze the complex environment. At the same time, I said that this capacity of projecting one's hypotheses on the world can become perverted and used to promote hate and sectarianism. The capacity to create worlds also presents a risk, but this time a social and ideological one. Individuals may be tempted to impose their "world" on other societies or groups.

This is the case for the fanaticism being unleashed in our times. This is also, more subtly, the case for democratic societies like France or new cultures that are calling into question the dominant republican model. This model constituted a shared world whose formidable capacity for absorption and consensus was described by the French writer André Chouraqui; it was also the world so poetically caricatured by the stormy alliance between the mayor and the priest in the films of Fernandel. But it lost its strength of conviction in the face of other cultures, other "worlds" whose values and structures are different.

It is to be hoped that conciliation will be reached by facing up to reality. But in *The Imaginary,* Jean-Paul Sartre rightly, in my opinion, points out that the world of the imaginary can be completely dissociated from reality: "Proust has shown well this abyss that separates the imaginary from the real, he has shown well that one can find no passage from one to the other and that the real is always accompanied by the collapse of the imaginary, even if there is no contradiction between the two, because the incompatibility comes from their nature and not from their content."[29] And further on: "The imaginative act is at once *constituting, isolating,* and *annihilating.*"[30] It "annihilates" the real. For Sartre, the dream fabricates an imaginary world, based on past experience and episodic memory, but completely independent of perception as it is unfurling (I use the word "fabricate" in the same sense as Philippe Descola in *La Fabrique des images* [Fabricating images], an exhibition he organized at the Quai Branly museum). Recent findings from neuroscience show the effective disconnection of all sensation during dreaming.

In short, vicariance is an essential process in our interactions with others. Strictly speaking, it simply describes the occasional replacement of a priest by his assistant. But the semantic context of the concept has expanded to describe the capacities of human thought, social conduct and behavior, and so forth. From a simple substitution of sensory receptors for perception, the meaning has evolved to include the capacity for finding diverse cognitive strategies for performing the same task, the use of several cognitive pathways to aid students' comprehension during learning, and the ability to feel others' emotions.

Looking at all of these various meanings in perspective makes clear that vicariance concerns a major property of life, and in particular, of the human brain: that of creating elegant solutions that are compatible with our beliefs, plans, and the state of our environment— in sum, our culture. These solutions take into account the astonishing diversity of life caused not only by the vagaries of genetics, but also by interaction with the environment over the course of epigenesis. Throughout evolution, living things have become endowed with an extraordinary flexibility and capacity for innovation. These solutions are adapted to our needs at any given moment, to the influence of our culture, and to our past experiences, and their diversity in turn induces the multiple manifestations of vicariance.

In other words, contrary to what we assumed at the outset, vicariance has not undergone any genuine metamorphosis. If this concept applies across a wide range of domains, it is not because it is ill defined, but rather because it embodies a fundamental property of life. Over the course of evolution, the constraint of self-contained worlds, as described by von Uexküll, gave way to a capacity, particularly well developed in humans, to invent possible "worlds" in the face of complex problems. The vicarious act is thus a creator of worlds by virtue of the creative tension between the particular and the universal. The cultural context of vicariance bears investigation.

In the words of linguists Daniel Haun and Christian Rapold: "In some domains, cultural diversity goes hand in hand with cognitive diversity, and a cross-cultural perspective should play a central part in understanding how variable adult cognition is built from a common cognitive foundation."[31] It is up to us to see to it that these very diverse "worlds" are harmonious and do not cause massacres among individuals living in different worlds, but rather, because of the virtues of vicariance, are a testament to the marvelous creative capacities of the human brain.

NOTES

ACKNOWLEDGMENTS

INDEX

Notes

Prologue

Epigraph: C. Ossola, "Le paradoxe herméneutique," in A. Berthoz, B. Stock, and C. Ossola, eds., *La Pluralité interpretative* (Paris: Collège de France, 2010), available at http://www.college-de-france.fr/. Quotation at end of epigraph from D. Hammarskjöld, *Markings*. "Only the hand that erases can write the true thing" is the epigraph to *Markings*.

1. A. Berthoz, *The Brain's Sense of Movement*, trans. G. Weiss (Cambridge, MA: Harvard University Press, 2000); *Emotion and Reason: The Cognitive Science of Decision Making*, trans. G. Weiss (Oxford: Oxford University Press, 2006); *Simplexity*, trans. G. Weiss (New Haven: Yale University Press, 2009). See also A. Berthoz and J.-L. Petit, *The Physiology and Phenomenology of Action*, trans. C. Macann (Oxford: Oxford University Press, 2008).
2. This observation was supplied by the philosopher Jean-Luc Petit, who was good enough to read my manuscript.

Introduction

1. A Web search reveals the many uses of the term "vicariance." For example, a boat that carries tourists to the Galapagos in Darwin's footsteps is named *Vicariance*.

2. Berthoz, *Simplexity.*

3. See http://www.etudes-litteraires.com/grammaire-conjonctions.php #ixzz227i-B3At8 and Samuel Bidaud, "Sur la notion linguistique de vicariance," *Onomázein,* 28 (December 2013): 29–41.

4. A. Prochiantz, *Qu'est-ce que le vivant?* (Paris: Seuil, 2012).

5. C. Hagège, *Halte à la mort des langues* (Paris: Odile Jacob, 2000).

6. J. Largeault, *Intuition et intuitionisme* (Paris: Vrin, 1993).

7. A. Rabatel, *La Construction textuelle du point de vue* (Lausanne: Delachaux and Niestlé, 1998); A. Rabatel, *Homo narrans* (Limoges: Lambert Lucas, 2008).

8. M. Lussault, *L'Avènement du monde* (Paris: Seuil, 2013), p. 280. See also M. Lussault, *L'Homme spatial* (Paris: Seuil, 2007); J. Lévy and M. Lussault, "Logique de l'espace. Esprit des Lieux," in J. Lévy, ed., *L'Invention du monde: une géographie de la mondialisation* (Paris: Presses de Sciences Po, 2008); J. Lévy, *Réinventer la France: trente cartes pour une nouvelle géographie* (Paris: Fayard, 2013).

9. E. Danblon, *L'Homme rhétorique: culture, raison, action* (Paris: Éditions du Cerf, 2013), p. 147.

10. The 6th Geneva Conference on Person-Centered Medicine was held on 29 April–1 May 2013 under the aegis of the World Health Organization and with the participation of twenty-eight international institutions and associations.

11. M. Bret, M.-H. Tramus, and A. Berthoz, "Interacting with an Intelligent Dancing Figure: Artistic Experiments at the Crossroads between Art and Cognitive Science," *Leonardo,* 38 (2005): 46–53.

12. M. Edwards, *Shakespeare et la comédie de l'émerveillement* (Paris: Desclée de Brouwer, 2003).

1. The Brain as a Problem-Solver

1. Groups of two or three letters in these words correspond to the first letters of the emperors: CAEsar, AUgustus, TIberius, CAligula, and so forth.

2. E. Mounier, *Traité du caractère* (Paris: Seuil, 1946), p. 721.

3. A. Binet and V. Henri, "La Psychologie individuelle," *L'Année psychologique,* 2 (1895): 415–465.

4. L. J. Cronbach, "Les Deux disciplines de la psychologie appliquée," *Revue de psychologie appliquée,* 8 (1957): 159–187. Details of the history of the relations between the two fields can be found in the work of

Jacques Lautrey: J. Lautrey, "Esquisse d'un modèle pluraliste du développement cognitif," in M. Reuchlin et al., eds., *Cognition: l'individuel et l'universel* (Paris: PUF, 1990), pp. 185–216; J. Lautrey, ed., *Universel et différentiel en psychologie* (Paris: PUF, 1995); J. Lautrey et al., in R. J. Sternberg et al., eds., *Models of Intelligence: International Perspectives* (Washington, DC: American Psychological Association, 2002).

5. M. Proust, *Remembrance of Things Past*, vol. 2: *The Guermantes Way and Cities of the Plain*, trans. C. K. S. Moncrieff and T. Kilmartin (New York: Random House, 1982), pp. 921, 649.

6. M. Reuchlin, "Processus vicariants et différences interindividuelles," *Journal de psychologie*, 2 (1978): 133–145. See also P.-Y. Gilles, *Psychologie différentielle* (Rosny-sous-Bois: Bréal, 1999), pp. 54–55.

7. J.-P. Sartre, *Search for a Method*, trans. H. E. Barnes (New York: Knopf, 1968), p. 100.

8. J. von Uexküll, *Mondes animaux et mondes humains* (Paris: Denoël, 1984); English translation: *A Foray into the Worlds of Animals and Humans*, trans. J. D. O'Neil (Minneapolis: University of Minnesota Press, 2010). The translation of *Umwelt* is controversial and has been alternatively rendered (in French, at any rate) as the equivalent of "worlds" (*mondes*) and "milieus." The proceedings of a symposium I organized with Y. Christen of the Fondation Ipsen reviews von Uexküll's ideas and the corresponding references to his work: A. Berthoz and Y. Christen, eds., *The Neurobiology of "Umwelt": How Living Beings Perceive the World* (Berlin: Springer, 2009).

9. Von Uexküll, *Mondes animaux et mondes humains*, pp. 59, 95.

10. H. Piéron, "Psychologie zoologique," in G. Dumas, *Nouveau traité de psychologie*, vol. 7 (Paris: PUF, 1941). He writes, "The world of perception (*Merkwelt*) is adjusted to the world of action (*Wirkwelt*) by 'perceptive characters' acting on the sensory receptors (*Wirkmalträger*) (indicators of action)." This is similar to von Uexküll's concept of "counterpoint."

11. Around 1915, Alfred Wegener was defending his theory of continental drift despite the opposition of the experts of the time. It would be more than fifty years before people admitted he was right.

12. I was given this example by Philippe Janvier, professor at the Museum National de l'Histoire Naturelle in Paris and a great expert in these questions. He generously helped with the editing of this section. See also the work of Gareth Nelson, at the University of Melbourne.

13. F. Ronquist, "Dispersal-Vicariance Analysis: A New Approach to the Quantification of Historical Biogeography," *Systematic Biology,* 46 (1997): 195–203.
14. Here the emphasis is on animal biogeography. But there is also a vast literature on vicariance in plant biogeography.

2. Perceiving and Acting

1. Some biogeographers question whether plate tectonics, the theory that explains continental drift, is responsible for the growth in species diversity, at least over the last two hundred million years.
2. Berthoz, *The Brain's Sense of Movement; Emotion and Reason; Simplexity;* and Berthoz and Petit, *The Physiology and Phenomenology of Action.*
3. See, for example, the article by J. M. Hillis et al., "Combining Sensory Information: Mandatory Fusion Within, but Not Between Senses," *Science,* 298 (2002): 1627–1630.
4. Corollary discharge is also sufficient for detecting the position of the eye. In 1974, Jacques Paillard showed that the constancy of focal precision during pursuit of a target is related to the postural system of reference, which provides yet more indirect information on the position of the eye. This polymodal parallelism comes into play at more or less complex levels of behavior.
5. G. Cioni and G. Sgandurra, "Normal Psychomotor Development," in O. Dulac et al., *Handbook of Clinical Neurology on Pediatric Neurology* (Amsterdam: Elsevier, 2013).
6. Moshé Feldenkrais was a physician who suffered a serious accident and subsequently constructed a method of rehabilitation inspired largely by his knowledge of physiology and biomechanics. This method is based on training body awareness and body image. See M. Feldenkrais, *Body Awareness as Healing Therapy: The Case of Nora* (Berkeley, CA: Frog Books, 1993); originally published as *La Conscience du corps* (Paris: Robert Laffont, 1971). His teachings have been promoted in France by Myriam Pfeiffer and her daughter Sabine, with whom I have organized many international meetings to foster dialog and debate with neuroscientists.
7. Feldenkrais, *Body Awareness as Healing Therapy,* p. 71 (italics mine).

8. F. de Waal, *Peacemaking among Primates* (Cambridge, MA: Harvard University Press, 1989).

9. See the work of Béatrice de Gelder and her colleagues on the physical expression of emotions.

10. See also G. Bollens, *Le Style des gestes: corporéité et kinésie dans le récit littéraire* (Lausanne: Éditions BHMS, 2008).

11. The Guardian Angels for a Smarter Life project of the European Union is a large initiative along these lines. See http://www.ga-project.eu/. On cyborgs, see, for example, S. Mann and H. Niedzviecki, *Cyborg: Digital Destiny and Human Possibility in the Age of the Wearable Computer* (Toronto: Doubleday, 2001). These ideas also inspired the film *Cyberman*.

12. J.-A. Sahel, S. Picaud, and A. B. Safran, "Vision artificielle: les prothèses rétiniennes," in J.-P. Changeux, ed., *L'Homme artificiel au service de la société* (Paris: Odile Jacob, 2007), pp. 191–209; A. P. Fornos et al., "Temporal Properties of Visual Perception on Electrical Stimulation of the Retina," *Investigative Ophthalmology and Visual Science,* 53 (2012): 2720–2731.

13. J. Rouger et al., "Evolution of Crossmodal Reorganization of the Voice Area in Cochlear-Implanted Deaf Patients," *Human Brain Mapping,* 33 (2012): 1929–1940; A. Amedi et al., "Shape Conveyed by Visual-Auditory Sensory Substitution Activates the Lateral Occipital Complex," *Nature Neuroscience,* 10 (2007): 687–689; M. Auvray, S. Hanneton, and J. K. O'Regan, "Learning to Perceive with a Visuo-Auditory Substitution System: Localisation and Object Recognition with 'The vOICe,'" *Perception,* 36 (2007): 416–430; M. J. Proulx et al., "Seeing 'Where' through the Ears: Effects of Learning-by-Doing and Long-Term Sensory Deprivation on Localization Based on Image-to-Sound Substitution," *PLoS One,* 3 (2008): e1840.

14. The CLONS project under the European Commission's Future and Emerging Technologies program. For otoliths, see: M. Demicolli et al., "Striola Magica: A Functional Explanation of Otolith Geometry," *Journal of Computational Neuroscience,* 35 (2013): 125–154.

15. The feelSpace project at the University of Osnabrück is designed to test the capacity of humans to use echolocation to orient themselves in space. See also the e-sense project at the University of Edinburgh.

16. P. Bach-y-Rita et al., "Vision Substitution by Tactile Image Projection," *Nature,* 221 (1969): 963–964; P. Bach-y-Rita and S. W. Kercel, "Sensory

Substitution and the Human-Machine Interface," *Trends in Cognitive Neuroscience*, 7 (2003): 541–546; P. Bach-y-Rita, *Brain Mechanisms in Sensory Substitution* (New York: Academic Press, 1972).

17. L. Renier and A. G. De Volder, "Cognitive and Brain Mechanisms in Sensory Substitution of Vision: A Contribution to the Study of Human Perception," *Journal of Integrative Neuroscience*, 4 (2005): 489–503; C. Poirier, A. G. De Volder, and C. Scheiber, "What Neuroimaging Tells Us about Sensory Substitution," *Neuroscience and Behavioral Reviews*, 31 (2007): 1064–1070.

18. K. A. Kaczmarek et al., "Electrotactile and Vibrotactile Displays for Sensory Substitution Systems," *IEEE Transactions Biomedical Engineering*, 38 (1991): 1–16; P. Bach-y-Rita et al., "Form Perception with a 49-Point Electrotactile Stimulus Array on the Tongue: A Technical Note," *Journal of Rehabilitative Research and Development*, 35 (1998): 427–430; M. Schurmann et al., "Touch Activates Human Auditory Cortex," *NeuroImage*, 30 (2006): 1325–1331; M. Tyler, Y. Danilov, and P. Bach-y-Rita, "Closing an Open-Loop Control System: Vestibular Substitution through the Tongue," *Journal of Integrative Neuroscience*, 2 (2003): 159–164.

19. J. P. Roll et al., "Inducing Any Virtual Two-Dimensional Movement in Humans by Applying Muscle Tendon Vibration," *Journal of Neurophysiology*, 101 (2009): 816–823.

20. E. Striem-Amit et al., "Reading with Sounds: Sensory Substitution Selectively Activates the Visual Word Form Area in the Blind," *Neuron*, 76 (2012): 640–652.

21. K. Jerbi et al., "Watching Brain TV and Playing Brain Ball, Exploring Novel BCI Strategies Using Real-Time Analysis of Human Intracranial Data," *International Review of Neurobiology*, 86 (2008): 159–168.

22. S. Dehaene and L. Cohen, "Two Mental Calculation Systems: A Case Study of Severe Acalculia with Preserved Approximation," *Neuropsychologia*, 29 (1991): 1045–1054.

23. L. Cohen, S. Dehaene, and P. Verstichel, "Number Words and Number Nonwords: A Case of Deep Dyslexia Extending to Arabic Numerals," *Brain*, 117 (1994): 267–279.

24. D. Kahneman, *Thinking, Fast and Slow* (New York: Farrar, Straus and Giroux, 2011).

25. O. Deroy and M. Auvray, "Reading the World through the Skin and Ears: A New Perspective on Sensory Substitution," *Frontiers in Psychology*, 3 (2012): 457.

26. J. R. Vidal et al., "Long-Distance Amplitude Correlations in the High γ Band Reveal Segregation and Integration within the Reading Network," *Journal of Neuroscience,* 9 (2012): 6421–6434.

27. See J. McIntyre et al., "The Brain as a Predictor: On Catching Flying Balls in Zero-G," in J. C. Buckey et al., *The Neurolab Spacelab Mission: Neuroscience Research in Space* (Houston: Lyndon B. Johnson Space Center, 2003), pp. 55–61; P. Senot et al., "When Up Is Down in Zero G: How Gravity Sensing Affects the Timing of Interceptive Actions," *Journal of Neuroscience,* 32 (2012): 1969–1973; P. Senot, S. Baillet, B. Renault, and A. Berthoz, "Cortical Dynamics of Anticipatory Mechanisms in Interception: A Neuromagnetic Study," *Journal of Cognitive Neuroscience,* 20 (2008): 1827–1838.

28. H. Hicheur et al., "Velocity and Curvature in Human Locomotion along Complex Curved Paths: A Comparison with Hand Movements," *Experimental Brain Research,* 162 (2005): 145–154.

29. I am citing from memory.

30. I described the properties of these internal models in *The Brain's Sense of Movement, Emotion and Reason,* and *Simplexity.*

31. H. Imamizu et al., "Human Cerebellar Activity Reflecting an Acquired Internal Model of a New Tool," *Nature,* 403 (2000): 192–195.

32. H. Wölfflin, "Prolegomena to a Psychology of Architecture," in H. F. Mallgrave and C. Contandriopoulos, eds., *Architectural Theory,* vol. 2, *An Anthology from 1871–2005* (Malden, MA: Blackwell, 2008), p. 75.

33. Olaf Blanke, a neurologist at Lausanne, and his group have published numerous reports on this subject: O. Blanke et al., "Stimulating Illusory Own-Body Perceptions," *Nature,* 419 (2002): 269–270; O. Blanke et al., "Out-of-Body Experience and Autoscopy of Neurological Origin," *Brain,* 127 (2004): 243–258; O. Blanke et al., "Linking Out-of-Body Experience and Self Processing to Mental Own-Body Imagery at the Temporoparietal Junction," *Journal of Neuroscience,* 25 (2005): 550–557.

34. A. Prochiantz (*Les Anatomies de la pensée* [Paris: Odile Jacob, 1997], p. 177ff.), proposed the interesting idea that the genetic organization of body modules, as it appears at the level of somites, is then echoed throughout the brain.

35. Berthoz, *Simplexity.*

36. I. Olive and A. Berthoz, "Combined Induction of Rubber-Hand Illusion and Out-of-Body Experiences," *Frontiers in Neuroscience,* 3 (2012): 128.

37. A. Maravita and A. Iriki, "Tools for the Body Schema," *Trends in Cognitive Science,* 8 (2004): 79–86. See also the articles from the 2012 Rome symposium: *Spatial Cognition and Embodiment.*
38. S. Bao, V. Chan, and M. Merzenich, "Cortical Remodelling Induced by Activity of Ventral Tegmental Dopamine Neurons," *Nature,* 412 (2001): 7983. The cortex receives very long projections from the dopaminergic neurons of the ventral tegmental area. These neurons are activated by novel stimuli or unexpected rewards. This dopamine activation probably reinforces cortical reorganization linked to learning. Another example is that of the primary auditory cortex, where dopamine release was observed during auditory learning, which modifies the representations of the frequencies of sounds in the cortex. Dopamine also modulates "long-term potential," that is, the increased sensitivity of neurons to specific messages, which is assumed to be a fundamental cellular mechanism in plasticity.
39. A. Michotte, *Causalité, permanence et réalité phénoménale* (Paris: Béatrice-Nauwelaerts, 1962), p. 97.
40. R. Grush and S. Kosslyn, "The Emulation Theory of Representation: Motor Control, Imagery, and Perception," *Behavioural and Brain Sciences,* 27 (2004): 377–442.
41. R. Llinás, *I of the Vortex: From Neurons to Self* (Cambridge, MA: MIT Press, 2002), p. 57.
42. Sartre, *Search for a Method,* p. 147. Italics mine.
43. J.-P. Vernant and M. Detienne, *Les Ruses de l'intelligence: la métis des Grecs* (Paris: Flammarion, 1974).
44. These observations are taken from various reports of the psychologist Théophile Ohlmann, mentioned earlier.
45. On coordination of head and limbs, see A. Berthoz and T. Pozzo, "Intermittent Head Stabilisation during Postural and Locomotory Tasks in Humans," in B. Amblard, A. Berthoz, and F. Clarac, eds., *Posture and Gait: Development, Adaptation, and Modulation* (Amsterdam: Elsevier, 1998), pp. 189–198. On similarity of laws, see S. Vieilledent et al., "Relationship between Velocity and Curvature of a Human Locomotor Trajectory," *Neuroscience Letters,* 305 (2001): 65–69; D. Bennequin et al., "Movement Timing and Invariance Arise from Several Geometries," *PLoS Computational Biology,* 5 (2009): e1000426. On gaze, see R. Grasso et al., "Development of Anticipatory Orienting Strategies during Locomotor Tasks in Children," *Neuroscience and Bio-*

behavioral Reviews, 22 (1997): 533–539; D. Bernardin et al., "Gaze Anticipation during Human Locomotion," *Experimental Brain Research,* 223 (2012): 65–78. On stereotypical trajectories, see Q. C. Pham, "Invariance of Locomotor Trajectories across Visual and Gait Direction Conditions," *Experimental Brain Research,* 210 (2011): 207–215.

46. J. L. Campos, "Imagined Self-Motion Differs from Perceived Self-Motion: Evidence from a Novel Continuous Pointing Method," *PLoS One,* 11 (2009): e7793.

47. C. Papaxanthis et al., "Comparison of Actual and Imagined Execution of Whole-Body Movements after a Long Exposure to Microgravity," *Neuroscience Letters,* 339 (2003): 41–44; K. Iseki et al., "Neural Mechanisms Involved in Mental Imagery and Observation of Gait," *NeuroImage,* 41 (2008): 1021–1031; S. Vieilledent et al., "Does Mental Simulation of Following a Path Improve Navigation Performance without Vision?" *Cognitive Brain Research,* 16 (2003): 238–249.

48. J. Wagner et al., "Mind the Bend: Cerebral Activations Associated with Mental Imagery of Walking along a Curved Path," *Experimental Brain Research,* 191 (2008): 247–255. Evidence has been found of activation in the putamen contralateral to the turn and in the caudate nucleus at initiation of turning, and of greater activation in the parahippocampus and the fusiform gyrus when walking along a curved path than along a straight line. Reports also document inactivation of the superior and medial gyrus involved in high-level integration of vestibular and multisensory information.

49. G. Ganis, W. L. Thompson, and S. M. Kosslyn, "Neuroimaging Evidence for Object Model Identification Theory," *Cognitive Brain Research,* 20 (2004): 226–241.

50. V. Moro et al., "Selective Deficit of Mental Visual Imagery with Intact Primary Visual Cortex and Visual Perception," *Cortex,* 44 (2008): 109–118.

51. D. Lehmann et al., "Core Networks for Visual-Concrete and Abstract Thought Content: A Brain Electric Microstate Analysis," *NeuroImage,* 49 (2010): 1073–1079.

52. W. Penfield and P. Perot, "The Brain Record of Auditory and Visual Experience: A Final Summary and Discussion," *Brain,* 86 (1963): 595–696; W. Penfield, "Vestibular Sensation and the Cerebral Cortex," *Annals of Otology, Rhinology, and Laryngology,* 66 (1957): 691–698; W. Penfield, "The Interpretive Cortex: The Stream of Consciousness in the

Human Brain Can Be Electrically Reactivated," *Science,* 129 (1959): 1719–1725. See also any of Penfield's works providing overviews of the field.

53. See D. Addis et al., "Episodic Simulation of Future Events Is Impaired in Mild Alzheimer's Disease," *Neuropsychologia,* 47 (2009): 2660–2671; D. L. Schacter et al., "Episodic Simulation of Future Events: Concepts, Data, and Applications," *Annals of the New York Academy of Sciences,* 1124 (2008): 39–60.

54. T. Suddendorf, "Mental Time Travel: Continuities and Discontinuities," *Trends in Cognitive Science,* 17 (2013): 151–152; T. Suddendorf and M. C. Corballis, "Behavioural Evidence for Mental Time Travel in Nonhuman Animals," *Behavioural Brain Research,* 31 (2010): 292–298.

55. D. L. Schacter, D. R. Addis, and R. L. Buckner, "Remembering the Past to Imagine the Future: The Prospective Brain," *Nature Review/Neuroscience,* 8 (2007): 657–661.

56. J. Summerfield, D. Hassabis, and E. Maguire, "Cortical Midline Involvement in Autobiographical Memory," *NeuroImage,* 44 (2009): 1188–1200.

57. E. Husserl, "Die Bernauer Manuscripte über das Zeitbewusstsein," *Husserliana,* 1917–1918, vol. 33, nos. 1, 2, ed. M. Fleischer (The Hague: M. Nijhoff, 1966). The English translation here is translated from J.-L. Petit's own translation into French.

58. D. L. Schacter and D. R. Addis, "The Ghosts of Past and Future," *Nature,* 445 (2007): 27.

59. A. Schnider et al., "Early Distinction between Memories That Pertain to Ongoing Reality and Memories That Do Not," *Cerebral Cortex,* 12 (2002): 54–61; A. Schnider, *The Confabulating Mind: How the Brain Creates Reality* (New York: Oxford University Press, 2008).

60. A. Viard et al., "Hippocampal Activation for Autobiographical Memories over the Entire Lifetime in Healthy Aged Subjects," *Cerebral Cortex,* 17 (2007): 2453–2467. See also the work by Pascale Piolino on autobiographical memory, in particular P. Piolino, B. Desgranges, and F. Eustache, *La Mémoire autobiographique: théorie et pratique* (Marseille: Solal, 2000).

61. S. Zeki, "The Disunity of Consciousness," *Trends in Cognitive Science,* 7 (2003): 214–218.

62. For more on this subject, see L. Naccache, *Le Nouvel Inconscient: Freud, Christophe Colomb des neurosciences* (Paris: Odile Jacob, 2006).

63. M. Menzocchi, E. L. Santarcangelo, C. G. Carli, and A. Berthoz, "Hypnotizability-Dependent Accuracy in the Reproduction of Haptically Explored Paths," *Experimental Brain Research*, 216 (2012): 217–223.

3. The Personal Body, Self, and Identity

1. For details of the discovery of mirror neurons, see G. Rizzolatti and S. Sinigaglia, *Les Neurones miroirs* (Paris: Odile Jacob, 2011).
2. M. Merleau-Ponty, *The Structure of Behavior*, trans. A. L. Fisher (Boston: Beacon, 1963), p. 148; Merleau-Ponty cites Goldstein, *Der Aufbau des Organismus Haag* (Leiden: Maartinus Nijhoff, 1934).
3. Michotte, *Causalité*, p. 124.
4. See, for example, S. Ferret, *L'Identité* (Paris: Flammarion, 1998); P. Ricoeur, *Oneself as Another*, trans. K. Blamey (Chicago: University of Chicago Press, 1992); C. Lévi-Strauss, *L'Identité* (Paris: Grasset, 1977); M. Mauss, "A Category of the Human Mind: The Notion of Person; the Notion of Self," in M. Carrithers et al., eds., *The Category of the Person* (Cambridge: Cambridge University Press, 1989); J.-C. Kaufmann, *L'Invention de soi: une théorie de l'identité* (Paris: Armand Colin, 2004); A. Ehrenberg, *La Fatigue d'être soi* (Paris: Odile Jacob, 1998); O. Houdé, *La Psychologie de l'enfant* (Paris: PUF, 2005); P. Rochat, *Le Monde des bébés* (Paris: Odile Jacob, 2006); J. Nadel and J. Decety, *Imiter pour découvrir l'humain* (Paris: PUF, 2005); S. J. Gould, *The Mismeasure of Man* (New York: W. W. Norton, 1996); J.-M. Monteil, *Soi et le context* (Paris: Armand Colin, 1993); M. Archer, *Being Human: The Problem of Agency* (Cambridge: Cambridge University Press, 2000); S. Gruszow, ed., *L'Identité: qui suis-je?* (Paris: Le Pommier, 2006).
5. D. Wiggins, in D. Carosella et al., *L'Identité changeante de l'individu* (Paris: L'Harmattan, 2008), p. 21.
6. T. Nathan, "À qui j'appartiens?" in Gruszow, ed., *L'Identité: qui suis-je?* pp. 30–52. Quotation is on p. 40.
7. J. Locke, *An Essay Concerning Human Understanding*, book 2, chap. 27, no. 9.
8. Ricoeur, *Oneself as Another*, pp. 143, 377.
9. Ibid., p. 14.
10. Ibid., p. 170.

11. S. Gallagher, "Philosophical Conceptions of the Self: Implications for Cognitive Sciences," *Trends in Cognitive Sciences*, 4 (2000): 14–21.

12. See the work of Harvard psychologist Daniel Schacter, cited in Chapter 1. B. Gaesser et al., "Imagining the Future: Evidence for a Hippocampal Contribution to Constructive Processing," *Hippocampus*, 23 (2013): 1150–1161.

13. U. Eco, *Baudolino*, trans. W. Weaver (New York: Harcourt, 2002).

14. The philosopher Jean-Luc Petit sent me his translation of Husserl's reasoning on intersubjectivity as the basis of unity and identity of the subject from appendix 27 of volume 39 of *Husserliana*. This text might provide one approach to explaining why dissociation of personality and identity occurs in so many forms. Ricoeur himself was unaware of this text (J.-L. Petit, pers. comm.), although he reinvented it in his own way in his "narrativist" critique of Derek Parfit. Here is Petit's version of the stages, based on Husserl's text: (1) The world (*Lebenswelt*) is the invariant of the congruent variety of my experiences; (2) a break in congruence leads to a duplicate world, whence a single subject becomes two according to their respective constitutional worlds; (3) for a solipsistic subject, nothing stands in the way of his duplication; (4) if others are my representations (theory of mind), the situation is unchanged; (5) if they are subjects with their own lives, then the mere fracture of congruence in my experience will lead to some discongruence: passing from world A to world B, I will be no less in their field of experience; (6) the account of their experience will contradict that of my own experience: my two worlds will be the worlds of a madman! Intersubjectivity masks the contradiction of the hypothesis of a duplicate subject. Conclusion: The unity and identity of the subject are the creation of intersubjectivity.

15. G. Simondon, *L'Individuation à la lumière des notions de forme et d'information* (Grenoble: Million, 2005), p. 31; G. Simondon, "The Position of the Problem of Ontogenesis," trans. G. Flanders, *Parrhesia*, no. 7 (2009): 4–16, citation from p. 10.

16. P. Descola, "Les Civilisations amérindiennes," in A. Cheng et al., *Les Grandes Civilisations* (Paris: Bayard, 2011), p. 243. See also P. Descola, *Par-delà nature et culture* (Paris: Gallimard, 2005).

17. E. H. Kantorowicz, *The King's Two Bodies: A Study in Mediaeval Political Theology* (Princeton, NJ: Princeton University Press, 1998), p. 7.

18. This symposium, co-organized with the occupational health psychologist Yves Clot, is available online at the Collège de France website. Following the symposium, a day-long discussion of the subject was organized by Jack Ralite in the French Senate.

19. T. Nathan, "Identity across Space and Time: Identity and Transnational Diasporas," in E. Ben-Rafael and Y. Sternberg, eds., *Transnationalism: Diasporas and the Advent of a New (Dis)order* (Boston: Brill, 2009), p. 188.

20. R. Harré, *Personal Being* (Oxford: Blackwell, 1983), p. 20.

21. *Paroles d'étoiles: Mémoires d'enfants cachés, 1939–1945* (Paris: Librio, 2002), p. 80. Emphasis mine.

22. M. Merleau-Ponty, *Phenomenology of Perception* (London: Routledge and Kegan Paul, 1962), p. 291.

23. These are called delusional misidentification syndromes and are frequently associated with schizophrenia.

24. Fregoli syndrome was discovered and described by Courbon and Fail in 1927. The following articles provide a comparative summary: H. D. Ellis, J. Whitley, and J.-P. Luauté, "Delusional Misidentification: The Three Original Papers on the Capgras, Frégoli, and Intermetamorphosis Delusions (Classic Text No. 17)," *History of Psychiatry*, 5 (17, part 1), (1994):117–146; N. M. J. Edelstyn, "Visual Processing in Patients with Fregoli Syndrome," *Cognitive Neuropsychiatry*, 1(2) (1996): 103–124. Some of this information was supplied by J. Arturo Silva, MD, Psychiatry Service, West Los Angeles VAMC, Los Angeles.

25. The expression comes from E. Couchot, *Les Technologies dans l'art* (Paris: Jacqueline Chambon, 1998), p. 13.

26. É. Pereny, ed., "Images interactives et jeux vidéo," *Questions théoriques*, 2012, p. 27. See also É. Pereny and É. A. Amato, "L'heuristique de l'avatar: polarités et fondamentaux des hypermédias et des cybermédias," *Revue des interactions médiatisées*, 11 (2010): 87–115.

27. É. A. Amato, *Le Jeu vidéo comme dispositif d'instanciation. Du phénomène ludique aux avatars en réseau*, Thesis, University of Paris 8, 2008. See also his many publications on this subject.

28. Pereny, "Images interactives et jeux vidéo," p. 136. On "delegation," see B. Rieder, *Métatechnologies et délégation: Pour un design orienté-société dans l'ère du Web 2.0*, Thesis, University of Paris 8, 2006.

29. Pereny, "Images interactives et jeux vidéo," p. 91.

30. The founder of humanoid robotics in Japan is Hirochika Inoue. Today, many major Japanese laboratories are working on these questions, such as those of Atsuo Takanishi at Waseda University in Tokyo, Yoshihiko Nakamura at the University of Tokyo, and Abderrahmane Keddar in Tsukuba. In France, the Aldebaran Society, headed by Bruno Maisonnier, has produced two humanoid robots, Nao et Roméo. At the LAAS in Toulouse, Jean-Paul Laumond is working with the Japanese HRP2 robot. Several other French teams have produced humanoids. Philippe Bidault heads ISIR at University of Paris 6, where research focuses on humanoids and multisensory interaction. In Germany, work is being done by Gordon Cheng in Munich and Thomas Mergner in Ulm, and in Pisa, Italy, Paolo Dario, who directed RoboCom, one of the pioneer projects in this area. In the United States, the work of Rodney Brooks is well known, and several teams, such as Boston Dynamics, are working on these problems. The Republic of Korea is also represented. This list is not exhaustive, and is intended merely to indicate the importance of the field, which holds its own conference every year.

31. My thanks to J.-P. Laumond for his help in editing this section on redundancy in robotics.

32. See the European Union project website on the RoboCom robot companion: www.robotcompanions.eu. Laumond's 2012 lectures at the Collège de France are also available in podcast on the website.

33. This is the robotics laboratory headed by Atsuo Takanishi. The collaboration takes place in the context of European projects coordinated by Paolo Dario at Pisa (the RoboCom project).

34. A comprehensive list of references on this issue can be found in A. P. Saygin et al., "The Thing That Should Not Be: Predictive Coding and the Uncanny Valley in Perceiving Human and Humanoid Robots Actions," *Social Cognitive and Affective Neuroscience,* 7 (2012): 413–422.

35. See, for example, R. Arkin, *Behavior Based Robotics* (Cambridge, MA: MIT Press, 1998).

36. This humanoid robot is installed in Toulouse, at the CNRS robotics laboratory (LAAS) as part of a Japanese collaboration. See O. Kanoun, J.-P. Laumond, and E. Yoshida, "Planning Foot Placements for a Humanoid Robot: A Problem of Inverse Kinematics," *International Journal of Robotics Research,* 30 (2011): 476–485, and J.-P. Laumond, "Simplexité et robotique: vers un génie de l'action encorporée," in J.-L. Petit and A.

Berthoz, eds., *Simplexité et complexité,* online collection of the Collège de France, 2013.

4. Vicariance and Changing Perspective

1. G. Galati, G. Pelle, A. Berthoz, and G. Committeri, "Multiple Reference Frames Used by the Human Brain for Spatial Perception and Memory," *Experimental Brain Research,* 206 (2010): 109–120.
2. Noa Eshkol was an Israeli dancer and choreographer (1924–2007). Together with her husband, Abraham Wachman, an architect, she created the Eshkol-Wachman movement notation system (1958).
3. E. Lobel et al., "Functional MRI of Galvanic Vestibular Stimulation," *Journal of Neurophysiology,* 80 (1998): 2699–2709; P. Kahane et al., "Reappraisal of the Human Vestibular Cortex by Cortical Electrical Stimulation Study," *Annals of Neurology,* 54 (2003): 615–624.
4. N. Tabareau et al., "Geometry of the Superior Colliculus Mapping and Efficient Oculomotor Computation," *Biological Cybernetics,* 97 (2007): 279–292.
5. J. Weiss et al., "Neural Consequences of Acting in Near versus Far Space: A Physiological Basis for Clinical Dissociations," *Brain,* 123 (2000): 2531–2541.
6. A. Berthoz, I. Viaud-Delmon, and S. Lambrey, "Spatial Memory during Navigation: What Is Being Stored, Maps or Movements?" in A. M. Galaburda, S. Kosslyn, and Y. Christen, eds., *The Languages of the Brain* (Cambridge, MA: Harvard University Press, 2002), pp. 288–306.
7. F. Yates, *The Art of Memory* (Chicago: University of Chicago Press, 1966); M. Carruthers, *The Book of Memory: A Study of Memory in Medieval Culture* (Cambridge: Cambridge University Press, 1990); M. Carruthers, *The Craft of Thought: Meditation, Rhetoric, and the Making of Images, 400–1200* (Cambridge: Cambridge University Press, 1998). See also B. Chatwin, *The Songlines* (New York: Viking, 1987); P. Rossi, *Logic and the Art of Memory,* trans. S. Clucas (Chicago: University of Chicago Press, 2000); L. Bolzoni, *The Gallery of Memory: Literary and Iconographic Models in the Age of the Printing Press,* trans. J. Parzen (Toronto: University of Toronto Press, 2001).
8. See A. Berthoz and R. Recht, eds., *Les Espaces de l'homme* (Paris: Odile Jacob, 2005).

9. There are Bayesian models of navigation. One study uses a unique approach: J. Diard, A. Berthoz, and P. Bessière, "Spatial Memory of Paths using Circular Probability Distributions: Theoretical Properties, Navigation Strategies, and Orientation Cue Combination," *Spatial Cognition and Computation: An Interdisciplinary Journal*, 13 (2013): 219–257.

10. Personal communication.

11. See also Berthoz and Petit, *The Physiology and Phenomenology of Action*.

12. M. Lafon, S. Tonnoir, A. Berthoz, and G. Thibault, "Analyse de la préparation et de la réalisation des tirs radiographiques pour le contrôle non destructif des soudures en CNPE," *Radioprotection*, 43 (2008): 409–428; G. Thibault et al., "How Does Horizontal and Vertical Navigation Influence Spatial Memory of Multifloored Environments?" *Attention, Perception, and Psychophysics*, 75 (2013): 10–15.

13. J. Barra et al., "Does an Oblique/Slanted Perspective during Virtual Navigation Engage Both Egocentric and Allocentric Brain Strategies?" *PLoS One*, 7 (2012): e49537.

14. I. Wiener, M. Lafon, and A. Berthoz, "Path Planning under Spatial Uncertainty," *Memory and Cognition*, 36 (2008): 495–504; A. H. Olivier et al., "Collision Avoidance between Two Walkers: Role-Dependent Strategies, *Gait Posture*, 38 (2013): 751–756; V. Belmonti et al., "Navigation Strategies as Revealed by Error Patterns on the Magic Carpet Test in Children with Cerebral Palsy," *Frontiers in Psychology*, 6 (2015): 880; V. Belmonti et al., "Cognitive Strategies for Locomotor Navigation in Normal Development and Cerebral Palsy," *Developmental Medicine and Child Neurology*, 57 suppl. s2 (2015): 31–36; V. Belmonti, G. Cioni, and A. Berthoz, "Switching from Reaching to Navigation: Differential Cognitive Strategies for Spatial Memory in Children and Adults," *Developmental Science*, 18 (2015): 569–586.

15. S. Lambrey and A. Berthoz, "Combination of Conflicting Visual and Nonvisual Information for Estimating Actively Performed Body Turns in Virtual Reality," *International Journal of Psychophysiology*, 50 (2003): 101–115.

16. D. Kimura, *Sex and Cognition* (Cambridge, MA: MIT Press, 1999); M. Hines, *Brain Gender* (New York: Oxford University Press, 2005). Note that a distinction is often made between sex (biology) and gender (culture).

17. I. Viaud-Delmon, I. Ivanenko, R. Jouvent, and A. Berthoz, "Sex, Lies, and Virtual Reality," *Nature Neuroscience*, 1 (2000): 15–16.

18. L. Cayhill, "Why Sex Matters for Neuroscience," *Nature Reviews Neuroscience*, 6 (2006): 477–484.

19. S. Lambrey and A. Berthoz, "Gender Differences in the Use of External Landmarks versus Spatial Representations Updated by Self-Motion," *Journal of Integrative Neuroscience*, 6 (2007): 1–23.

20. S. Springer and G. Deutsch, *Left Brain, Right Brain: Perspectives from Cognitive Neuroscience*, 5th ed. (Cranbury, NJ: W. H. Freeman, 2001).

21. S. Lambrey et al., "Distinct Visual Perspective-Taking Strategies Involve the Left and Right Medial Temporal Lobe Structures Differently," *Brain*, 31 (2008): 523–534.

22. L. Rondi-Reig et al., "Impaired Sequential Egocentric and Allocentric Memories in Forebrain-Specific-NMDA Receptor Knock-Out Mice during a New Task Dissociating Strategies of Navigation," *Journal of Neuroscience*, 26 (2006): 4071–4081; K. Iglói et al., "Sequential Egocentric Strategy Is Acquired as Early as Allocentric Strategy: Parallel Acquisition of These Two Navigation Strategies," *Hippocampus*, 19 (2009): 1199–1211; K. Iglói et al., "Lateralized Human Hippocampal Activity Predicts Navigation Based on Sequence or Place Memory," *Proceedings of the National Academy of Sciences*, 107 (2010): 14466–14471.

23. E. Burguière et al., "Spatial Navigation Impairment in Mice Lacking Cerebellar LTD: A Motor Adaptation Deficit?" *Nature Neuroscience*, 8 (2005): 1292–1294.

24. In humans, the crossed relation between the cerebellum and the hippocampus was shown by K. Iglói et al., "Interaction between Hippocampus and Cerebellum Crus I in Sequence-Based but not Place-Based Navigation," *Cerebral Cortex*, 25 (2015): 4146–4154.

25. I discussed this in *Simplexity*.

26. H. Poincaré, *The Foundations of Science: Science and Hypothesis, The Value of Science, Science and Method*, trans. G. B. Halsted (New York: Science Press, 1913), pp. 421–422.

27. M. Wraga et al., "Imagined Rotations of Self versus Objects: An fMRI Study," *Neuropsychologia*, 43 (2005): 1351–1361.

28. M. A. Amorim et al., "Updating an Object's Orientation and Location during Nonvisual Navigation: A Comparison between Two Processing Modes," *Perception and Psychophysics*, 59 (1997): 404–418; A. Berthoz et al., "Dissociation between Distance and Direction during Locomotor

Navigation," in R. G. Golledge, ed., *Wayfinding Behavior: Cognitive Mapping and Other Spatial Processes* (Baltimore: Johns Hopkins University Press, 1999), pp. 328–348.

29. Lambrey et al., "Distinct Visual Perspective-Taking Strategies."
30. S. Ortigue et al., "Pure Representational Neglect after Right Thalamic Lesion," *Annals of Neurology*, 3 (2001): 401–404.
31. Berthoz, *Emotion and Reason*. See also *Nature*, 500 (2013): 521.
32. A. Collins and É. Koechlin, "Reasoning, Learning, and Creativity: Frontal Lobe Function and Human Decision-Making," *PLoS Biology*, 10 (2012): e1001293.

5. The Stages of Vicariance

1. This idea has long been the subject of debates that are beyond the scope of this book.
2. D. A. Fair et al., "Functional Brain Networks Develop from a 'Local to Distributed' Organization," *PLoS Computational Biology*, 5 (2009): e1000381.
3. S. Dehaene and J.-P. Changeux, "Experimental and Theoretical Approaches to Conscious Processing," *Neuron*, 28 (2011): 200–227.
4. M. H. Johnson, "Functional Brain Development in Humans," *Nature Reviews Neuroscience*, 2 (2001): 475–483; M. H. Johnson and D. Mareschal, "Cognitive and Perceptual Development during Infancy," *Current Opinion in Neurobiology*, 11 (2001): 213–218.
5. See the references to the work of the group of Olivier Houdé in N. Poirel et al., "Number Conservation Is Related to Children's Prefontal Inhibitory Control," *PLoS One*, 7 (2012): e40802; A. Lubin et al., "Numerical Transcoding Proficiency in 10-Year-Old Children Is Associated with Gray Matter Interindividual Differences: A Voxel-Based Morphometry Study," *Frontiers in Psychology*, 4 (2013): 197; S. Rossi et al., "Structural Brain Correlates of Executive Engagement in Working Memory: Children's Interindividual Differences Are Reflected in the Anterior Insular Cortex," *Neuropsychologia*, 51 (2013): 1145–1150.
6. I give examples of this substitution in my previous work, as in the case of the neural basis of gaze control.
7. A. Cachia et al., "The Shape of the ACC Contributes to Cognitive Efficiency in Preschoolers," *Journal of Cognitive Neuroscience*, 26 (2014): 96–106. See O. Houdé, *Le Raisonnement* (Paris: PUF, 2014).

8. T. K. Hensch, "Critical Period Plasticity in Local Cortical Circuits," *Nature Reviews Neuroscience*, 6 (2005): 877–888; S. Sugiyama et al., "Experience-Dependent Transfer of Otx2 Homeoprotein into the Visual Cortex Activates Postnatal Plasticity," *Cell*, 8 (2008): 508–520.

9. P. Rochat, "Five Levels of Self-Awareness: As They Unfold Early in Life," *Consciousness and Cognition*, 12 (2003): 717–731. The psychologist Axel Cleeremans distinguishes three levels of self-awareness: L. Legrain, A. Cleeremans, and L. Destrebecqz, "Distinguishing Three Levels in Explicit Self-Awareness," *Consciousness and Cognition*, 20 (2011): 578–585.

10. P. D. Zelazo, J. A. Sommerville, and S. Nichols, "Age-Related Changes in Children's Use of External Representations," *Developmental Psychology*, 35 (1999): 1059–1071; P. D. Zelazo, "The Development of Conscious Control in Childhood," *Trends in Cognitive Science*, 8 (2004): 12–17.

11. J. Redshaw and T. Suddendorf, "Foresight beyond the Very Next Event: Four-Year-Olds Can Link Past and Deferred Future Episodes," *Frontiers in Psychology*, 9 (2013): 404.

12. D. J. Povinelli et al., "Development of Young Children's Understanding That the Recent Past Is Causally Bound to the Present," *Developmental Psychology*, 35 (1999): 1426–1439.

13. J. Bastin et al., "Timing of Posterior Parahippocampal Gyrus Activity Reveals Multiple Scene Processing Stages," *Human Brain Mapping*, 34 (2012): 1357–1370.

14. Povinelli et al., "Development of Young Children's Understanding."

15. J. Bullens et al., "Developmental Time Course of the Acquisition of Sequential Egocentric and Allocentric Navigation Strategies," *Journal of Experimental Psychology*, 107 (2010): 337–350; N. Poirel et al., "Evidence of Different Developmental Trajectories for Length Estimation according to Egocentric and Allocentric Viewpoints in Children and Adults," *Experimental Psychology*, 58 (2011): 142–146.

16. F. Joly and A. Berthoz, eds., *Julian Ajuriaguerra. Développement corporel et relations avec autrui. Actes du colloque d'hommage à Julian* (Neuilly-Plaisance: Éditions du Papyrus, 2013).

17. A. Berthoz, "Contrôle cognitif des fonctions sensori-motrices et vicariance: Quelques hypothèses pour la rééducation et la remédiation," in Joly and Berthoz, eds., *Julian Ajuriaguerra*, pp. 223–234.

18. See, for example, Berthoz, Stock, and Ossala, eds., *La Pluralité inter-prétative;* and A. Berthoz, "Le Changement de point de vue," in F. Gros, F. Terré, and B. Tournafond, eds., *Être humain* (Paris: CNRS Éditions, 2014), pp. 57–72.

19. A. Berthoz, *The Brain's Sense of Movement;* Berthoz, Stock, and Ossala, eds., *La Pluralité interprétative;* A. Berthoz, *Les Fondements cognitifs de la tolérance* (in press).

6. Vicariance and Brain Plasticity

1. H. Morishita and T. K. Hensch, "Critical Period Revisited: Impact on Vision," *Current Opinion in Neurobiology,* 18 (2008): 101–107.

2. A. A. Brewer, "Visual Maps: To Merge or Not to Merge," *Current Biology,* 19 (2009): R945–R947.

3. J. C. Dahmen, D. Hartley, and A. J. King, "Stimulus-Timing-Dependent Plasticity of Cortical Frequency Representation," *Journal of Neuroscience,* 28 (2008): 13629–13639.

4. R. C. Froemke, M. M. Merzenich, and C. E. Schreiner, "A Synaptic Memory Trace for Cortical Receptive Field Plasticity," *Nature,* 450 (2007): 425–429; J. F. Hunzinger, V. H. Chan, and R. C. Froemke, "Learning Complex Temporal Patterns with Resource-Dependent Spike Timing-Dependent Plasticity," *Journal of Neurophysiology,* 108 (2012): 551–566.

5. See the work of Nicoletta Berardi and Lamberto Maffei's team at Pisa. For example, S. Baldini et al., "Enriched Early Life Experiences Reduce Adult Anxiety-Like Behavior in Rats: A Role for Insulin-Like Growth Factor 1," *Journal of Neuroscience,* 33 (2013): 11715–11723; L. Baroncelli et al., "Enriched Experience and Recovery from Amblyopia in Adult Rats: Impact of Motor, Social, and Sensory Components," *Neuropharmacology,* 62 (2012): 2388–2397.

6. F. McNab et al., "Changes in Cortical Dopamine D1 Receptor Binding Associated with Cognitive Training," *Science,* 323 (2009): 800–802.

7. K. Woollett and E. A. Maguire, "Navigational Expertise May Compromise Anterograde Associative Memory," *Neuropsychologia,* 47 (2009): 1088–1095; K. Woollett and E. A. Maguire, "Exploring Anterograde Associative Memory in London Taxi Drivers," *Neuroreport,* 23 (2012): 885–888.

8. C. Xerri, "Plasticity of Cortical Maps: Multiple Triggers for Adaptive Reorganization Following Brain Damage and Spinal Cord Injury," *Neuroscientist*, 18 (2012): 133–148.

9. J. O'Shea et al., "Functionally Specific Reorganization in Human Premotor Cortex," *Neuron*, 54 (2007): 479–490.

10. For more on the subject, see L. Cohen et al., "Learning to Read without a Left Occipital Lobe: Right-Hemispheric Shift of Visual Word Form Area," *Annals of Neurology*, 56 (2004): 890–894; F. Vargha-Khadem, E. Isaacs, and V. Muter, "A Review of Cognitive Outcome after Unilateral Lesions Sustained during Childhood," *Journal of Child Neurology*, 9 (1994) Suppl. 2: 67–73; L. Hertz-Pannier et al., "Late Plasticity for Language in a Child's Non-Dominant Hemisphere: A Pre- and Post-Surgery fMRI Study," *Brain*, 125 (2002): 361–372; R. A. Muller et al., "Language Organization in Patients with Early and Late Left-Hemisphere Lesion: A PET Study," *Neuropsychologia*, 37 (1999): 545–557; F. Liegeois et al., "Language Reorganization in Children with Early-Onset Lesions of the Left Hemisphere: An fMRI Study," *Brain*, 127 (2004): 1229–1236; B. A. Shaywitz et al., "Disruption of Posterior Brain Systems for Reading in Children with Developmental Dyslexia," *Biological Psychiatry*, 52 (2002): 101–110; G. Di Virgilio and S. Clarke, "Direct Interhemispheric Visual Input to Human Speech Areas," *Human Brain Mapping*, 5 (1997): 347–354.

11. R. Leiguarda et al., "Globus Pallidus Internus Firing Rate Modification after Motor-Imagination in Three Parkinson's Disease Patients," *Journal of Neural Transmission*, 116 (2009): 451–455.

12. Various definitions of "remediation" can be found. In France, psychiatrists have created a cognitive remediation society (Association Francophone de Remédiation Cognitive). See also J. D. Bolter and R. Grusin, *Remediation: Understanding New Media* (Cambridge, MA: MIT Press, 2000), in which "remediation" denotes evolving media. The term thus has multiple meanings.

13. In *The Elusive Obvious* (Capitola, CA: Meta Publications, 1981, pp. 75–76), Moshé Feldenkrais writes: "I first used the individual manipulative technique Functional Integration and the group technique Awareness through Movement during World War II. Already at that time I worked on one side of the body exclusively during a whole lesson. The other side remained passive or motionless all through the lesson. I

wanted to create the greatest possible sensory contrast in the nervous structures and also facilitate awareness of the differences kinesthetically. I thought that the different organization of one side of the cortex and the corresponding side of the body would slowly diffuse to the other side. . . . What would be transferred to the other side of the brain is the (learned) better patterns, by [the person's] own feeling and judgment."

14. L. Mottron and M. Dawson, "The Autistic Spectrum," *Handbook of Clinical Neurology,* 111 (2013): 263–271.

15. N. Georgieff and M. Speranza, eds., *Psychopathologie de l'intersubjectivité* (Paris: Elsevier-Masson, 2013).

7. Sympathy and Empathy

1. In conjunction with my colleagues Carlo Ossola and Brian Stock, I organized an interdisciplinary symposium on this subject at the Collège de France that brought together experts in hermeneutics and religion—anthropologists, psychologists, neurophysiologists, and legal scholars: see Berthoz, Ossola, and Stock, eds., *La Pluralité interpretative.* Also worth mentioning is a recent symposium at the Collège de France organized by legal expert Mireille Delmas-Marty titled "Hominisation et humanisation."

2. C. Frith and D. Wolpert, eds., *The Neuroscience of Social Interaction* (New York: Oxford University Press, 2004), p. 274. See also the commentary by J. Barreqsi and C. Moore, "Intentional Relations and Social Understanding," *Nature Neuroscience,* 7 (2004): 9.

3. H. Kober et al., "Functional Grouping and Cortical-Subcortical Interactions in Emotion: A Meta-Analysis of Neuroimaging Studies," *NeuroImage,* 42 (2008): 998–1031.

4. N. Sebanz, G. Knoblich, and W. Prinz, "Representing Others' Actions: Just Like One's Own?" *Cognition,* 88 (2003): B11–B21; D. Kourtis, N. Sebanz, and G. Knoblich, "Predictive Representation of Other People's Actions in Joint Action Planning: An EEG Study," *Social Neuroscience,* 8 (2013): 31–42.

5. N. Ramnari and C. Miall, "A System in the Human Brain for Predicting the Actions of Others," *Nature Neuroscience,* 7 (2004): 185–190.

6. D. Wenke et al., "What Is Shared in Joint Action? Issues of Co-Representation, Response Conflict, and Agent Identification," *Review of Philosophy and Psychology,* 2 (2011): 147–172.

7. See the even-handed historical review of these concepts in *Une histoire de l'empathie* (Paris: Odile Jacob, 2012) by Jacques Hochmann, another review by Serge Tisseron, *L'Empathie au coeur du jeu social* (Paris: Albin Michel, 2010), and an outline of a theory of empathy in a collection that I edited with Gérard Jorland, *L'Empathie* (Paris: Odile Jacob, 2004). See also review articles by the philosopher Bérangère Thirioux: G. Jorland and B. Thirioux, "Note sur l'origine de l'empathie," *Revue de métaphysique et morale,* 2 (2008): 269–280; B. Thirioux and A. Berthoz, "Physiology of Empathy: How Intersubjectivity Is the Correlate of Objectivity," in J. Aden, T. Grimshaw, and H. Penz, eds., *Teaching Language and Culture in an Area of Complexity* (Brussels: Peter Lang, 2011), pp. 45–60. Finally, for an interdisciplinary debate on empathy, see the proceedings of a symposium organized by the Centre Culturel International de Cerisy in 2011.

8. A. Barliya et al., "Expression of Emotion in the Kinematics of Locomotion," *Experimental Brain Research,* 225 (2013): 159–176. The references in this article contain most of the citations relevant to the subject.

9. This text by C. Rogers and M. Kinget, in *Psychothérapies et relations humaines* (Paris: Béatrice-Nauwelaerts, 1962) is cited by Hochmann, *Une Histoire de l'empathie,* p. 113.

10. G. Dezecache, "Evidence for Unintentional Emotional Contagion beyond Dyads," *PLoS One,* 8 (2013): e67371.

11. G. Rizzolatti et al., "Resonance Behaviors and Mirror Neurons," *Archives italiennes de biologie,* 137 (1999): 85–100; G. Buccino et al., "Action Observation Activates Premotor and Parietal Areas in a Somatotopic Manner: An fMRI Study," *European Journal of Neuroscience,* 13 (2001): 400–404; V. Gallese, C. Keysers, and G. Rizzolatti, "A Unifying View of the Basis of Social Cognition," *Trends in Cognitive Science,* 8 (2004): 396–403; M. Fabbri-Destro and G. Rizzolatti, "Mirror Neurons and Mirror Systems in Monkeys and Humans," *Physiology (Bethesda),* 23 (2008): 171–179.

12. M. Iacoboni et al., "Cortical Mechanisms of Human Imitation," *Science,* 286 (1999): 2526–2528.

13. C. Keysers et al., "A Touching Sight: SII/PV Activation during the Observation and Experience of Touch," *Neuron,* 42 (2004): 335–346.

14. U. Hasson et al., "Intersubject Synchronization of Cortical Activity during Natural Vision," *Science,* 303 (2004): 1634–1640.

15. See the sources cited in note 7, above, for the history of the concept of empathy. For the etymology of *Einfühlung,* see G. Jorland and B. Thirioux, "Note sur l'origine de l'empathie," *Revue de métaphysique et de morale,* April 2008, pp. 269–280.

16. See Berthoz, *Emotion and Reason.*

17. Jean Decety, a French psychologist now living in the United States, is an expert on empathy and changes of perspective in relations with others. In addition to his many scientific articles, see his books on the subject: J. Decety and W. J. Ickes, *The Social Neuroscience of Empathy* (Cambridge, MA: MIT Press, 2009); J. Decety, *Empathy: From Bench to Bedside* (Cambridge, MA: MIT Press, 2011). See also F. de Vignemont and T. Singer, "The Empathic Brain: How, When, and Why?" *Trends in Cognitive Science,* 10 (2006): 435–441. For more on the subject, for example, on the idea of "social gaze," see the work of K. Vogely and G. R. Fink: L. Schilbach et al., "Eyes on Me: An fMRI Study of the Effects of Social Gaze," *Social Cognitive and Affective Neuroscience,* 6 (2011): 393–403. See also the work of S. Berthoz and J. Grèzes cited below.

18. E. Lobel et al., "Functional MRI of Galvanic Vestibular Stimulation," *Journal of Neurophysiology,* 80 (1998): 2699–2709; P. Kahane, D. Hoffmann, L. Minotti, and A. Berthoz, "Reappraisal of the Human Vestibular Cortex by Cortical Electrical Stimulation Study," *Annals of Neurology,* 54 (2003): 615–624; C. Lopez, O. Blanke, and F. Mast, "The Human Vestibular Cortex Revealed by Coordinate-Based Activation Likelihood Estimation Meta-analysis," *Journal of Neuroscience,* 212 (2012): 159–179.

19. M. Merleau-Ponty, *Signs,* trans. R. C. McCleary (Evanston, IL: Northwestern University Press, 1964), p. 169.

20. K. Vogeley and G. R. Fink, "Neural Correlates of First Person-Perspective," *Trends in Cognitive Science,* 7 (2003): 431–437.

21. Research on empathic processes is very active and has produced a range of findings. See the reports from the group of Tania Singer: H. G. Engen and T. Singer, "Empathy Circuits," *Current Opinion in Neurobiology,* 23 (2013): 275–282; B. C. Bernhardt and T. Singer, "The Neural Basis of Empathy," *Annual Reviews of Neuroscience,* 35 (2012): 1–23.

22. Klimecki et al., "Differential Pattern of Functional Brain Plasticity after Compassion and Empathy Training," *Social Cognitive and Affective Neuroscience,* 9 (2013): 873–879.

23. B. Thirioux et al., "Walking on a Line: A Motor Paradigm using Rotation and Reflection Symmetry to Study Mental Body Transformations," *Brain and Cognition,* 70 (2009): 191–200.
24. J. B. Bavelas, "Form and Function in Motor Mimicry: Topographic Evidence That the Primary Function Is Communicative," *Human Communication Research,* 14 (1988): 275–300.
25. B. Thirioux et al., "Mental Imagery of Self-Location during Spontaneous Self-Other Interactions in Standing and Moving Humans: An Electrical Neuroimaging Study," *Journal of Neuroscience,* 30 (2010): 7202–7214.
26. B. Thirioux et al., "Cortical Dynamics of Empathy and Sympathy: When and How Do We Select between the Two Behaviors during a Social Interaction?" (Submitted.)
27. Tisseron, *L'Empathie.*

8. Vicarious Emotion

1. On pain networks, see the recent work of O. M. Klimecki, "Functional Neural Plasticity and Associated Changes in Positive Affect after Compassion Training," *Cerebral Cortex,* 23 (2013): 1552–1561. For a philosophical approach to the topic, see V. Gallese and P. Jacob, "Vicarious Pain: Imagination, Mirroring, or Perception," *Frontiers of Neurosciences,* 5 (2011): 27.
2. K. W. Saakvitne, L. A. Pearlman, and the Staff of the Traumatic Stress Institute, *Transforming the Pain: A Workbook on Vicarious Traumatization* (New York: W. W. Norton, 1996).
3. D. S. Sade, "Sociometrics of *Macaca mulatta.* I. Linkages and Cliques in Grooming Matrices," *Folia Primatologica,* 18 (1972): 196–223, quotation from p. 209.
4. N. I. Eisenberger, M. D. Lieberman, and K. D. Williams, "Does Rejection Hurt? An fMRI Study of Social Exclusion," *Science,* 302 (2003): 290–292; C. L. Masten, S. A. Morelli, and N. I. Eisenberger, "An fMRI Investigation of Empathy for 'Social Pain' and Subsequent Prosocial Behavior," *NeuroImage,* 55 (2011): 381–388.
5. S. C. Welten, M. Zeelenberg, and S. M. Breugelmans, "Vicarious Shame," *Cognition and Emotion,* 26 (2012): 836–846.
6. P. Michel et al., "Neurobiological Underpinnings of Shame and Guilt: A Pilot fMRI Study," *Social Cognitive and Affective Neuroscience,* 9 (2012): 150–157.

7. J. Grèzes, S. Berthoz, and R. E. Passingham, "Amygdala Activation When One Is the Target of Deceit: Did He Lie to You or to Someone Else?" *NeuroImage,* 30 (2006): 601–608.
8. S. Berthoz et al., "Affective Response to One's Own Moral Violations," *NeuroImage,* 31 (2006): 945–950.
9. I. Pouga, S. Berthoz, B. de Gelder, and J. Grèzes, "Individual Differences in Socioaffective Skills Influence the Neural Bases of Fear Processing: The Case of Alexithymia," *Human Brain Mapping,* 31 (2010): 1469–1481.
10. N. I. Eisenberger, "The Pain of Social Disconnection: Examining the Shared Neural Underpinnings of Physical and Social Pain," *Nature Reviews Neuroscience,* 13 (2003): 421–434. Eisenberger has done extensive work on the subject of social pain and the differences between the sexes: see, e.g., N. I. Eisenberger et al., "An fMRI Study of Cytokine-Induced Depressed Mood and Social Pain: The Role of Sex Differences," *NeuroImage,* 47 (2009): 881–890.

9. Vicarious Learning

1. Y. Clot, *La Fonction psychologique du travail* (Paris: PUF, 2006); Y. Clot and J. Leplat, "La Méthode clinique en ergonomie et en psychologie du travail," *Le Travail humain,* 68 (2005): 289–316; Y. Clot, *Le Travail sans l'homme?* (Paris: Éditions La Découverte, 2008); Y. Clot, *Travail et pouvoir d'agir* (Paris: PUF, 2008).
2. J. Grégoire and T. Lubart, "Hauts Potentiels des enfants: force ou faiblesse? Identifier leurs aptitudes pour développer leurs talents," *ANAE,* 25, no. 119 (2012).
3. From a review by Michel Huteau of a book by Maurice Reuchlin (*Évolution de la psychologie différentielle,* Paris: PUF, 1999), in *Raison présente,* winter 2001, no. 137. See also M. Huteau, *Style cognitif et personnalité: la dépendance/indépendance à l'égard du champ* (Lille: Presses Universitaires de Lille, 1987).
4. The work of Théophile Ohlmann provides a comprehensive account of the ideas and findings of this school of thought: T. Ohlmann, "Évocabilité différentielle des référentiels spatiaux, posture et orientation spatiale," in V. Nougier and J.-P. Bianchi, eds., *Pratiques sportives et modélisation du geste* (Grenoble: Université Joseph-Fourier, 1990), pp. 215–240; T. Ohlmann, "La Perception de la verticale lors de con-

flits vision/posture: un exemple de processus vicariants," in M. Reuchlin et al., eds., *Connaître différemment* (Nancy: Presses Universitaires de Nancy, 1990), pp. 33–66; T. Ohlmann, "Les Systèmes perceptifs vicariants," in M. Reuchlin et al., eds., *Cognition: l'individuel et l'universel* (Paris: PUF, 1990), pp. 21–58; T. Ohlmann and C. Marendaz, "Vicariances et affordances: deux outils pour l'ergonomie cognitive," *Science et defense,* 91 (1991): 372–391; T. Ohlmann and C. Marendaz, "Vicarious Processes Involved in Spatial Perception," in S. Wapner and J. Demick, eds., *Field Dependence-Independence: Bio-Psycho-Social Factors across the Life Span* (Hillsdale, NJ: Lawrence Erlbaum, 1991), pp. 106–129.

5. A. Carter, L. T. Connor, and A. W. Dromerick, "Rehabilitation after Stroke: Current State of the Science," *Neuroscience Reports,* 10 (2010): 3158–3166. See also J. Hidler et al., "Multicenter Randomized Clinical Trial Evaluating the Effectiveness of the Lokomat in Subacute Stroke," *Neurorehabilitation and Neural Repair,* 23 (2009): 5–13.

6. M. Abdallah-Pretceille, "L'École, l'Université, la Recherche face à la diversité culturelle," in Katérina Stenou, ed., *Déclaration universelle de l'Unesco sur la diversité culturelle: Commentaires et propositions* (Paris: Unesco, 2013), pp. 14–22, quotation on p. 20.

7. Learning by vicarious reinforcement is an immense field of research that is beyond the scope of this book.

8. The notion of fictive learning methods in fact encompasses several pedagogical approaches.

9. M. Liljeholm, C. J. Molloy, and J. P. O'Doherty, "Dissociable Brain Systems Mediate Vicarious Learning of Stimulus Response and Action Outcome Contingencies," *Journal of Neuroscience,* 23 (2012): 9878–9886.

10. A. Bandura, *L'Apprentissage social* (Brussels: Mardaga, 1995); A. Bandura, *Auto-efficacité: le sentiment d'efficacité personnelle* (Brussels: De Boeck, 2007). On Bandura, see B. Guerrin, "Albert Bandura and His Work," *Recherche en soins infirmiers,* 108 (2012): 106–116.

11. I often describe the situation thus: "In the United States a young person is considered as having something to offer, and in France, as being in need of teaching!"

12. For summaries of Vygotsky's ideas, see works by Nathalie Bulle and Julien Gauthier.

13. N. Bulle, "Série L. Vygotski, 2: de Piaget à Vygotski," *Revue skolè.fr,* March 2009, available at http://skhole.fr/2-de-piaget-à-vygotski.

14. A. Tryphon and J. Vonèche, eds., *Piaget-Vygotsky: The Social Genesis of Thought* (Hove, East Sussex, UK: Psychology Press, 1996).

15. Educated in philosophy, animal and plant biology, and literature, Antoine de La Garanderie taught philosophy and general culture in many major secondary institutions in and around Paris and chaired conferences at the Faculty of Letters at the Catholic Institute of Paris. He served as director of the Audiovisual Institute of Paris, was awarded an honorary professorship at the Catholic University of Paris and the Catholic University of the West, and directed research at the University of Lyon II. He won a prize from the Académie française for *La Valeur de l'ennui* (1968). See, among other works, *Les Profils pédagogiques: discerner les aptitudes scolaires* (Paris: Bayard, 1980).

16. J.-F. Billeter, *Leçons sur Tchouang Tseu* (Paris: Allia, 2006), p. 61. This book contains the lectures given by the author at the Collège de France.

17. R. B. Welch, B. Bridgeman, J. A. Williams, and R. Semmler, "Dual Adaptation and Adaptive Generalization of the Human Vestibulo-Ocular Reflex," *Perception and Psychophysics,* 60 (1998): 1415–1425; H. A. Cunningham and R. B. Welch, "Multiple Concurrent Visual-Motor Mappings: Implications for Models of Adaptation," *Journal of Experimental Psychology: Human Perception and Performance,* 20 (1994): 987–999; M. Shelhamer, A. Aboukhalil, and R. Clendaniel, "Context Specific Adaptation of Saccade Gain Is Enhanced with Rest Intervals between Changes in Context State," *Annals of the New York Academy of Sciences,* 1039 (2005): 166–175; S. Lambrey, I. Viaud-Delmon, and A. Berthoz, "Influence of a Sensory Conflict on the Memorization of a Path Traveled in Virtual Reality," *Cognitive Brain Research,* 14 (2002): 177–186; A. Berthoz, I. Viaud-Delmon, and S. Lambrey, "Spatial Memory during Navigation: What Is Being Stored, Maps or Movements?"

18. I. Dumontheil, P. Panagiotaki, and A. Berthoz, "Dual Adaptation to Sensory Conflicts during Whole-Body Rotations," *Brain Research,* 1072 (2006): 119–132.

19. M. Lafon, M. Vidal, and A. Berthoz, "Selective Influence of Prior Allocentric Knowledge on the Kinesthetic Learning of a Path," *Experimental Brain Research,* 194 (2009): 541–552.

20. D. Bennequin et al., "Movement Timing and Invariance Arise from Several Geometries," *PLoS Computational Biology,* 25 (2009): e1000426.

21. Berthoz, *The Brain's Sense of Movement; Emotion and Reason.*

22. There is a rich literature on the neural basis of ocular saccades and the relationship between attention and gaze displacement. I have touched on certain aspects of this area in previous work, where I emphasized the fundamental role of inhibitory circuits in triggering saccades. This inhibition allows the brain both to modulate and to select relevant movements. In *The Brain's Sense of Movement* I also discussed the remarkable ability of saccades to substitute for the vestibulo-ocular reflex. This substitution is an excellent example of vicarious functioning. Recently, we studied these mechanisms using intracranial recordings in people with epilepsy. See S. Freyermuth et al., "Neural Basis of Saccadic Decision Making in the Human Cortex," in S. Funahashi, ed., *Representation and Brain* (Tokyo: Springer, 2008), pp. 199–216; Vidal et al., "Long-Distance Amplitude Correlations."

Epilogue

1. From a personal conversation with Gabriel Ruget, mathematician and former director of the École normale supérieure.
2. Sartre, *Search for a Method*, p. 49.
3. Ibid., p. 78.
4. P. van Andel and D. Bourcier, *De la sérendipité dans la science, la technique et le droit: leçons de l'inattendu* (Chambéry: L'Act Mem, 2009). On the role of chance, see also N. N. Taleb, *The Black Swan: The Impact of the Highly Improbable* (New York: Random House, 2007).
5. See the comprehensive list of definitions of metaphor on Cédric Detienne's website. See also P. Fontanier, *Les Figures du discours* (Paris: Flammarion, 1977), p. 100; G. Genette, "Métonymie chez Proust," *Figures III* (Paris: Seuil, 1972); "La lexicalisation de la métaphore," in M. Le Guern, *Sémantique de la métaphore et de la métonymie* (Paris: Larousse, 1973); N. Charbonnel and G. Kleiber, eds., *La Métaphore entre philosophie et rhétorique* (Paris: PUF, 1999); D. Sperber and D. Wilson, "Ressemblance et communication," in D. Andler, ed., *Introduction aux sciences cognitives* (Paris: Gallimard, 2004); and M. Fumaroli, *Le Livre des métaphores* (Paris: Robert Laffont, 2012).
6. G. Lakoff and M. Johnson, *Les Métaphores dans la vie quotidienne*, trans. Michel de Fornel and Jean-Jacques Lecercle (Paris: Minuit, 1985); G. Lakoff, "Explaining Embodied Cognition Results," *Topics in Cognitive Science*, 4 (2012): 773–785.

7. B. Bowdle and D. Gentner, "The Career of Metaphor," *Psychological Reviews,* 112 (2005): 193–216.

8. T. T. Kircher et al., "Neural Correlates of Metaphor Processing in Schizophrenia," *NeuroImage,* 34 (2007): 281–289; C. F. Norbury, "The Relationship between Theory of Mind and Metaphor: Evidence from Children with Language Impairment and Autistic Spectrum Disorder," *British Journal of Developmental Psychology,* 23 (2005): 383–399.

9. G. Le Dorze and J.-L. Nespoulous, "Anomia in Moderate Aphasia: Problems in Accessing the Lexical Representation," *Brain and Language,* 37 (1985): 381–400.

10. D. Anaki, M. Faust, and S. Kravetz, "Cerebral Hemispheric Asymmetries in Processing Lexical Metaphors," *Neuropsychologia,* 36 (1998): P353–P362; G. L. Schmidt, C. J. DeBuse, and C. A. Seger, "Right Hemisphere Metaphor Processing? Characterizing the Lateralization of Semantic Processes," 100 (2007): 127–141; G. Schmidt et al., "Beyond Laterality: A Critical Assessment of Research on the Neural Basis of Metaphor," *Journal of the International Neuropsychological Society,* 16 (2010): 1–5; J. Yang, "The Role of the Right Hemisphere in Metaphor Comprehension: A Meta-Analysis of Functional Magnetic Resonance Imaging Studies," *Human Brain Mapping,* 35 (2014): 107–122; G. Bottini et al., "The Role of the Right Hemisphere in the Interpretation of Figurative Aspects of Language: A Positron Emission Tomography Activation Study," *Brain,* 117 (1994): 1241–1253; Y. Arzouan et al., "Dynamics of Hemispheric Activity during Metaphor Comprehension: Electrophysiological Measures," *NeuroImage,* 36 (2007): 222–231.

11. E. Chen, P. Widick, and A. Chatterjee, "Functional-Anatomical Organization of Predicate Metaphor Processing," *Brain and Language,* 107 (2008): 194–202; R. W. Gibbs, "Metaphor Interpretation as Embodied Simulation," *Mind and Language,* 21 (2006): 434–458.

12. J. Kable, L. Lease-Spellmeyer, and A. Chatterjee, "Neural Substrates of Action Event Knowledge," *Journal of Cognitive Neuroscience,* 14 (2002): 795–805; A. Damasio and D. Tranel, "Nouns and Verbs Are Retrieved with Differently Distributed Neural Systems," *Proceedings of the National Academy of Sciences of the United States of America,* 90 (1993): 4957–4960.

13. D. Sperber and D. Wilson, "Loose Talk," in S. Davis, ed., *Pragmatics: A Reader* (Oxford: Oxford University Press, 1991), pp. 540–549, quotation on p. 549; D. Sperber and D. Wilson, "A Deflationary Account of Meta-

phors," *The Cambridge Handbook of Metaphor and Thought* (New York: Cambridge University Press, 2008), pp. 84–105, quotation on p. 103.

14. See especially M. Laganaro, "Production et compréhension des metaphors chez l'enfant: de la similitude à la métaphore," *Archives de psychologie*, Université de Genève, Éditions Médecine et Hygiène, 253 (1997): 141–165; L. J. Stites and S. Özçalişkan, "Developmental Changes in Children's Comprehension and Explanation of Spatial Metaphors for Time," *Journal of Child Language*, 40 (2013): 1123–1137.

15. Y. Clot, "Le garçon de bloc: étude d'ethnopsychologie du travail," *Éducation permanente*, 116 (1993): 97–109.

16. Y. Clot, "Le problème des catachrèses en psychologie du travail: un cadre d'analyse," *Le Travail humain*, 60 (1997): 113–119. In his article Clot cites the French and Russian schools of occupational psychology, for example: V. De Keyser, "Communications sociales et charge mentale dans les postes automatisés," *Psychologie française*, 28 (1983): 239–245; A. Savoyant, "Éléments d'un cadre d'analyse de l'activité: quelques conceptions essentielles de la psychologie soviétique," *Cahiers de psychologie*, 22 (1979): 17–28; D. Ochanine, "Le rôle des images opératives dans la régulation des activités de travail," *Psychologie et éducation*, 3 (1978): 63–72; L. S. Vygotski, *Pensée et langage*, trans. F. Sève (Paris: Messidor, 1985 [1934]).

17. D. Faïta et al., *Signer la ligne: les aspects humains de la conduite des trains*, Report by the SNCF and CCE/SNCF, 1996.

18. P. Rabardel, *Les Hommes et les technologies: approche cognitive des instruments contemporains* (Paris: Armand Colin, 1995), p. 119.

19. Descola, "Les Civilisations amérindiennes," p. 216.

20. Berthoz, Ossola, and Stock, eds., *La Pluralité interpretative*.

21. See G. Bouchard et al., *L'Utopie* (Montreal: Presses de l'Université de Montréal, 1985); Lucian, of Samosata, et al., *Voyages au pays de nulle part* (Paris: Robert Laffont, 1990); *Millénarisme et utopie dans les pays anglo-saxons* (Toulouse: Presses Universitaires du Mirail, 1998).

22. L. Holberg, *Niels Klim's Journey under the Ground*, trans. J. Gierlow (Boston: Saxton, Peirce & Co., 1845).

23. J. Verne, *Jules Verne's Magellania*, trans. B. Irvy (New York: Welcome Rain Publishers, 2002).

24. Danblon, *L'Homme rhétorique*, p. 124.

25. Ibid., p. 125; see also p. 167. Italics are Danblon's, except for the last expression.

26. J.-F. Billeter, *Un Paradigme* (Paris: Allia, 2012), p. 30.
27. Ibid., p. 29.
28. Ibid., p. 30.
29. J.-P. Sartre, *The Imaginary: A Phenomenological Psychology of the Imagination,* trans. J. Webber (London: Routledge, 2004).
30. Ibid., p. 146.
31. D.-B. Haun and C.-J. Rapold, "Variation in Memory for Body Movements across Cultures," *Current Biology,* 19 (2009): R1068-R1069. The quotation is from p. 1069. See also S. C. Levinson, *Space in Language and Cognition: Explorations in Cognitive Diversity* (Cambridge: Cambridge University Press, 2003); D.-B. Haun et al., "Cognitive Cladistics and Cultural Override in Hominid Spatial Cognition," *Proceedings of the National Academy of Sciences of the United States of America,* 103 (2006): 17568–17573.

Acknowledgments

I thank Odile Jacob, who, yet again, was good enough to let me take the risk of publishing an essay on a subject at the disciplinary frontier. I wish I could have suggested to her a term somewhat easier on the ear than "vicariance." But its currency as well as my inability to find a novel alternative for this marvelous property of life argued for keeping it. Odile kept careful watch to ensure that the text was both scientific and accessible to the broadest possible readership. This is a challenge she helps all of her authors to surmount through her vast knowledge and her implacable kindness.

Thanks, too, to Émilie Barian for agreeing to read and correct the first draft of the book. Nicolas Witkowski kindly shared with me the benefit of his substantial experience in taking on the main editing of the book. He advised me with an appealing mix of firmness and tolerance, and with remarks that were always deeply perceptive and suggested in a way that left me feeling the decisions were mine.

I would also like to extend my gratitude to the friends and colleagues who corrected or contributed to parts or all of the manuscript: Daniel Bennequin, Vincent Hayward, Philippe Janvier, Étienne Koechlin, Jean-Paul Laumond, Jean-Luc Petit, Jean Petitot, Alain Prochiantz, and Brian Stock.

Thank you to Maya Berthoz, who was a partner in this enterprise through her selflessly shared remarks, suggestions, and encouragement; her judgment is more precious to me than I can say.

Finally, my thanks to the entire team at Odile Jacob Press for their professionalism and patience with my delays and many errors. Without them, this work would still be a future project; with them, it became a book.

Last but not least I would like to express my deepest gratitude to Giselle Weiss, who agreed to translate another one of my books into English. She offered for this one her exceptional competence, intuition of the right expression, and most generous work and efforts. I had little to correct from her initial text, so great was her capacity to find the adequate way to reflect my thinking. Her combination of rigor and flexiblity in finding the best way to transform the complex French way of saying things into English scientific language is amazing. Any inadequate expression is my fault. Thank you, Giselle!

Index